U0215625

INTERNATIONAL CREATIVE RESTAURANT DESIGN

欧陆

II

深圳视界文化传播有限公司 编

国际创意
餐厅设计

中国、印度尼西亚
日本、加拿大、
罗马尼亚、希腊
……

中国林业出版社
China Forestry Publishing House

PROFILE | Einstein and Associates

Born in Jakarta – Indonesia Leo Einstein Franciscus graduated from the University of Pelita Harapan in 2008, he received his Bachelor Degree cum laude in Architecture. In 2006 he got his first internship at Shie Fen architect, one of legendary Indonesian architect. Upon graduated from university, Einstein worked at Atelier Cosmas Gozali and Willis Kusuma Architect from January 2009 – December 2012 as Project Designer & Senior Project Designer. Some of the F&B project at Jakarta he involved when his working time such as Ocha & Bella, SKYE, and MOOVINA. In 2013, he finally had his courage to found a boutique architecture interior design firm, Einstein and Associates to create distinct projects. A series of recognized works with sensuous and unique characteristic has born afterwards. Einstein & Associates Project Wilshire SCBD, Lemongrass,Bottega Ristorante Mega Kuningan, Wilshire Senopati, 3rd Avenue, Djati Drinkery, Populi Kitchen & Bar, Seroeni St. Moritz Nam Thai Kitchen & Bar, La Fusion, Bottega SCBD, Dirty Laundry, So Thai by Chandara Plaza Indonesia, Zhuma Senayan City, Ground Makassar. Einstein & Associates does very concern with details to inspire and indulge. This firm has established a new paradigm in the industry, encompassing a multitude of disciplines with creating thoughtful and integrated design process, focusing on creating transformative experiences. From the works that explained in this book, show that Einstein & Associates like to play with finishes and materials; from many kinds of wood, metal, brass, copper, leather, mosaic tiles, fabric, and many more. The ultimate key to our design is the balance, like yin and yang. When Einstein founded the firm, He already has a clear picture of what he wants in term of design. Einstein & Associates plans to break through different boundaries. We want to make sure that we are not staying idly within our comfort zone and doing the same old stuff. We want to stay creative and hungry for innovations while looking for new opportunities.

PREFACE

Luxury is a condition of great comfort, ease, elegance, wealth and timeless. It speaks of the privilege and exclusivity. It is also the enjoyment of the best in life: the experience of beauty, knowledge, and humanity at their deepest and most inspiring. From this perspective, luxury can remind us to love the life we're living, and not to simply live it, not let this fantastic existence go to waste. Beauty must be sought out, and one must make the conscious choice to discover quality and to enjoy it. Luxury is a term that I constantly use to describe my work with attention to the details. Luxury and timeless work which includes eye-catching designs always in my signature style to be. I used materials that portray my luxury and timeless design like mosaic brass, copper, stone, wood and some plants. Those are the materials that I like, and I think that they will never go out of style. For me, in creating luxury design, it must be able to represent exclusivity, something extraordinary, and also timeless. Sometimes, it has to be sophisticated and be something not predictable. I want people can experience the uniqueness the things that I created. Luxury has three important aspects, which is quality, excellent aesthetic and visual charm and the last but not least is people's perception. Luxury has a lot to do with taste, and therefore it is very subjective. The three aspects are closely related to the brand and lifestyle.

I was in charge of several restaurant projects in Jakarta, like Ocha & Bella, Skye, and Moovina while I was working as a senior designer at Willis Kusuma Architects, then I opened my design firm in 2013. And now Einstein & Associates was in charge of several restaurant projects like Wilshire SCBD, Lemongrass, Wilshire Senopati, 3rd Avenue, Djati Drinkery, La Fusion, Bottega SCBD, Dirty Laundry, Ground Makassar, etc.

I define luxury in terms of the quality of experience. I guess market offers a great diversity in terms of that kind of experience because everybody is much more informed about design and travelling and now instagramable design is the trends of the millennial society. The criteria in making the luxurious design for me not only that it has many expensive finishes, but also how we put it together. If we can combine the contrast and finishes very well, it will make a luxurious design. When I started my design, I also see things of real beauty that can be achieved with the wealth of experience, difference spaces and combine beautiful material which rich in colors that match with the interior. In my perspective, one of the important things is the details. Details can be more powerful than words. It's a 'wow factor' that goes beyond expectations and can be translated into many different ways. It can create a unique magical experience and ambience of the area. The other things are materials. It can help to create environments; they provide character to each space, identity and make them unique.

My advice for the reader is when seeing a new project, don't see them as what they are now, but see on perspective what we could make them as a luxurious space and design. Last but not least, luxury is often criticized, and it should not be forgotten.

Einstein and Associates
Leo Einstein Franciscus

CONTENTS
目录
欧陆食代 II

THE ROMANCE OF TASTE BUDS SURROUNDED BY FLOWERS

繁花簇拥的味蕾浪漫

Project Name | 项目名称
LA FUSION

Design Company | 设计公司
EINSTEIN & ASSOCIATES

Designer | 设计师
LEO EINSTEIN FRANCISCUS

Area | 项目面积
435 ㎡

La Fusion is a fusion restaurant that combines French and Asian food. The idea behind this restaurant is to blend two different cultures, tropical Asia with contemporary Europe, to form a beautiful modern piece of architecture and interior work that also perfectly embodies Pekanbaru as a trading city. The designer infused a French art deco visual style into the design, making it a modern tropical local haven with art deco touch, consistent with the whole fusion idea of the restaurant. There are two important key points from the design that the designer wants to convey, the first one is the fusion on every aspect of the restaurant, whether the food, the materials or the building and its natural environment. The second one is to emphasize Pekanbaru's attractive aspects like its nature and weather, its exposure to the outside world as a trading port and last but not least its openness to many different ethnicities.

As a tropical city rich with exotic and unique materials that distinguish Pekanbaru from the rest, it is only logical that the designer decided to use local materials like local wood and natural stones.

LA FUSION

TOILET

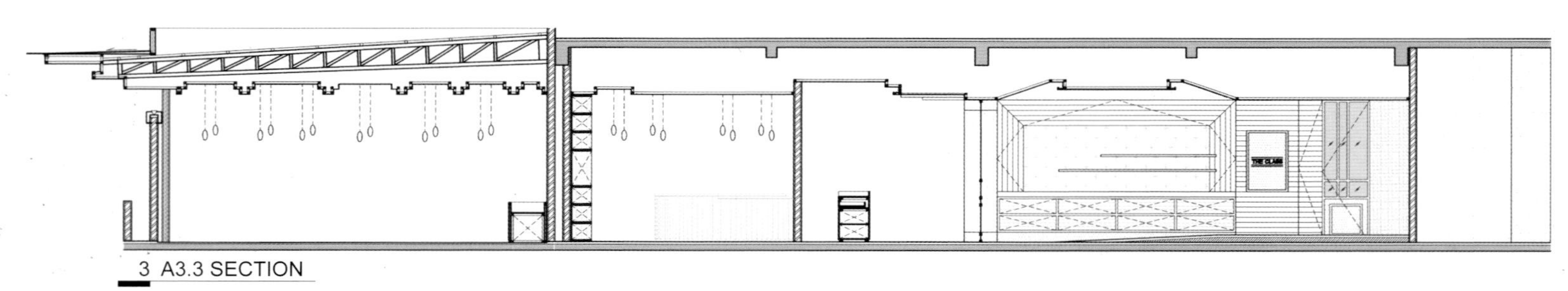

3 A3.3 SECTION

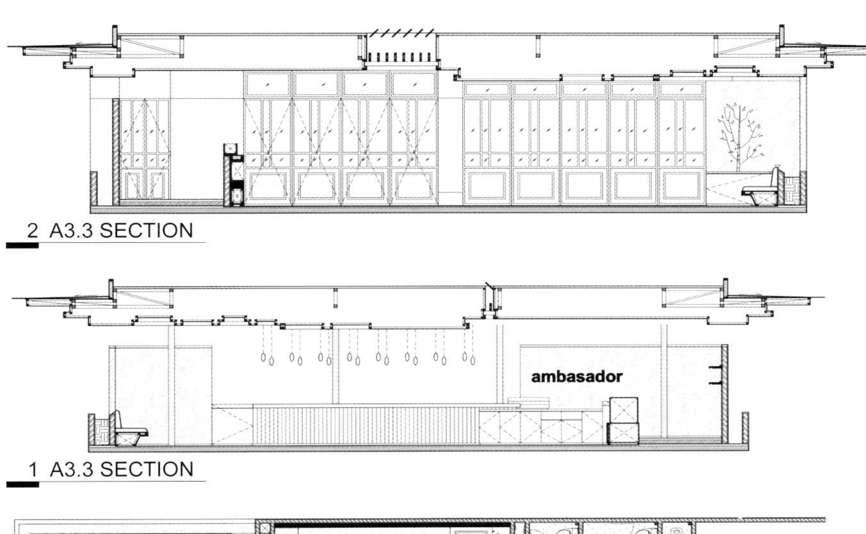

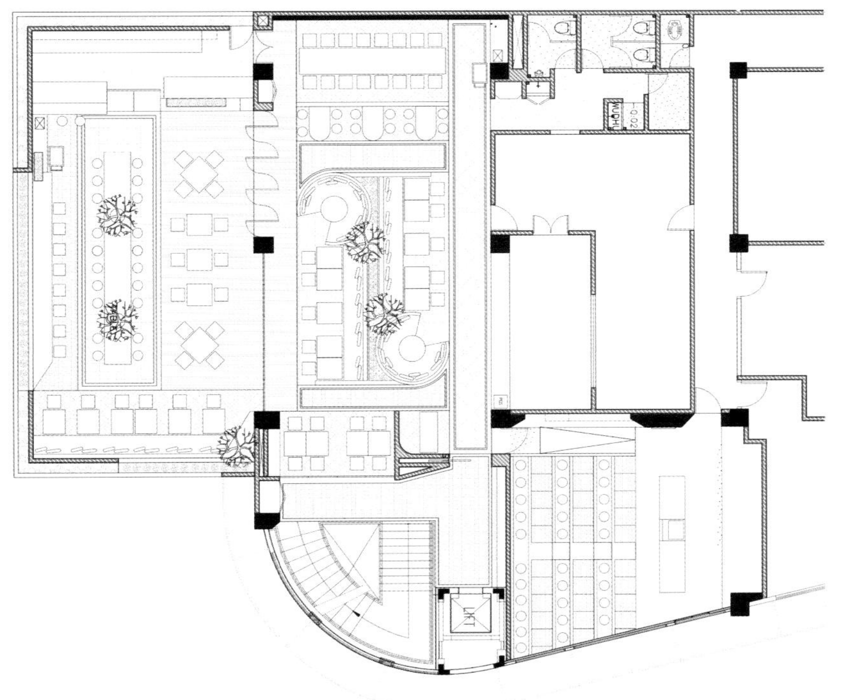

The fusion of modern tropical style and art deco visual style can be seen from the use of uniformed and geometrical shapes, art deco styled doors and windows, the chevron wooden floor, the black and white woven pattern on the dining chair, gold filmed glass and mirrors combined with the local wood and natural stones. In a mission to break the barriers between the indoor, outdoor area and the beautiful nature of Pekanbaru, the designer deliberately placed plants like big trees and floral murals, used the color blue symbolizing the bright sky as the dominant color, painted ceilings with floral themed images, a skylight, stunning floral mosaic tiles as an accent of the restaurant.

The restaurant is divided into two main areas, the indoor and the outdoor area. It also has a unique facility, a classroom named "The Class" dedicated to educating kids who are interested in the baking process of cakes, bread and pastry. The indoor area consists of the open kitchen, the center dining area and a VIP room at the back. While the outdoor area consists of the open bar and the dining area.

La Fusion是一家融合法国和亚洲食物的混合型餐厅。这家餐厅背后的理念是结合热带亚洲和现代欧洲这两种不同的文化，从而形成一个美丽的现代化的建筑和室内作品，也完美地体现北干巴鲁这一座贸易城市。设计师将法国装饰艺术的视觉风格注入到设计中，使之成为一个具有装饰风格的现代热带本土港口，与整个餐厅的融合理念相一致。设计师传达两个要点，一是餐厅各个方面的融合，无论是食物、材料还是建筑以及其自然环境。二是强调北干巴鲁的优势方面，除了自然环境和气候，还有以其贸易港口的身份为外界所熟知并且对不同种族的包容性。

作为一个拥有丰富独特原材料的富有异国情调的热带城市，这是北干巴鲁与其他城市的区别所在。所以设计师决定使用当地的木材和天然石头等材料，这是合乎逻辑的。

该案例将统一的几何形状、富有装饰风格的门窗、人字纹地板、餐椅上的黑白编织图案、金色的玻璃和镜子与当地的木材和天然石材结合在一起，可以体现出现代热带风格与装饰艺术视觉风格的融合。为了打破室内、室外和北干巴鲁美丽的大自然之间的屏障，设计师刻意摆放高大树木的植物和花卉壁画，用蓝色作为主色调象征蓝天，用以花卉为主题的图片、天窗和花卉印花瓷砖来彩绘天花板，将其作为餐厅的亮点。

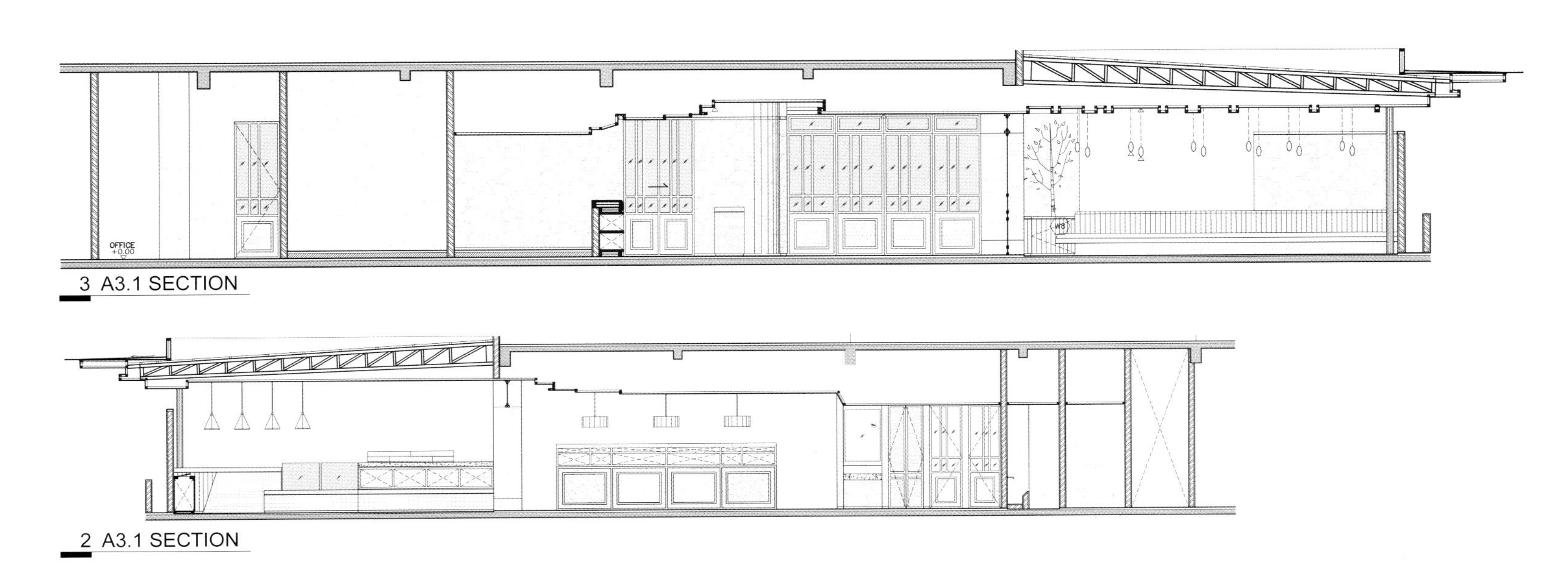

3 A3.1 SECTION

2 A3.1 SECTION

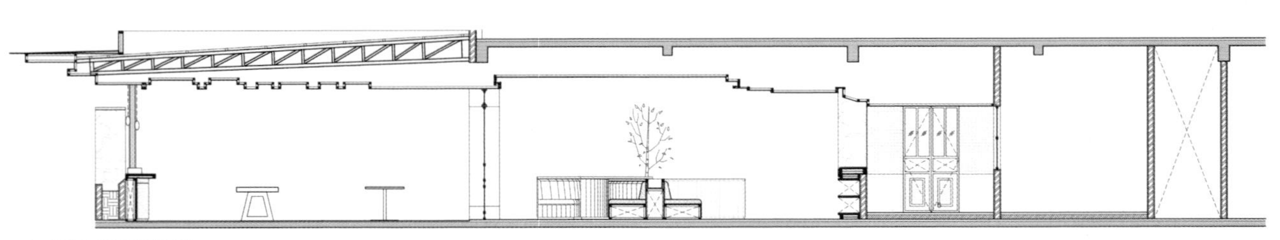

1 A3.1 SECTION

The experience of entering the restaurant is filled with the domination of natural stones covering almost all of the restaurant's walls with light blue windows as an accent. The long wooden and concrete ceiling from the indoor to the outdoor serves as the connecting thread between the two areas. Feeling invited by the playfulness of the floor materials, people are unconsciously enticed to enter the VIP room, entering the room then immediately noticing the long communal table, accommodating up to sixteen people made from green mosaic tiles and framed with brass in the VIP room, with crystal hanging lights enhancing the atmosphere the room. Brass framed floral mosaic tiles placed at the center of the VIP room as a focal point.

The indoor and outdoor area is only separated by the Art Deco styled doors and windows, with firm lines on the wooden ceiling and concrete modular ceiling. Dominated with wood and natural stones, the outdoor area is filled with the color blue, to give a sense of freshness and relaxation and serves as a canvas to mirror the beautiful weather of Pekanbaru, with colorful furniture as splashes of colors complementing the blue sky, just like flowers under a clear blue sky. In the outdoor dining area, the open bar also has the communal bar table surrounding the edge of the building, for people who want to enjoy the day or night view of Pekanbaru. There is also a triangle skylight in the middle of the outdoor area to add more sunlight and air circulation. All in all, La Fusion is a place where fusion happens.

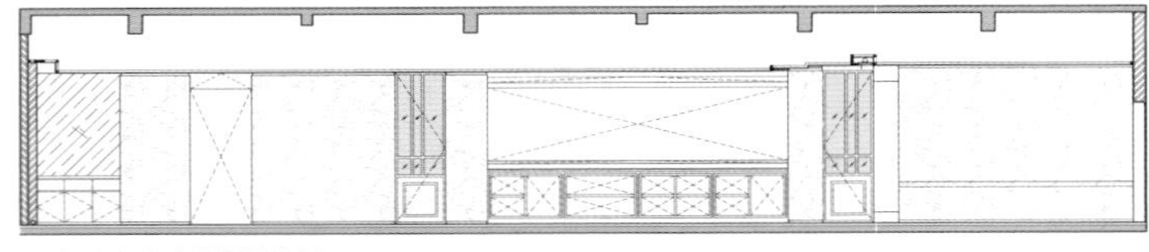

3 A3.2 SECTION

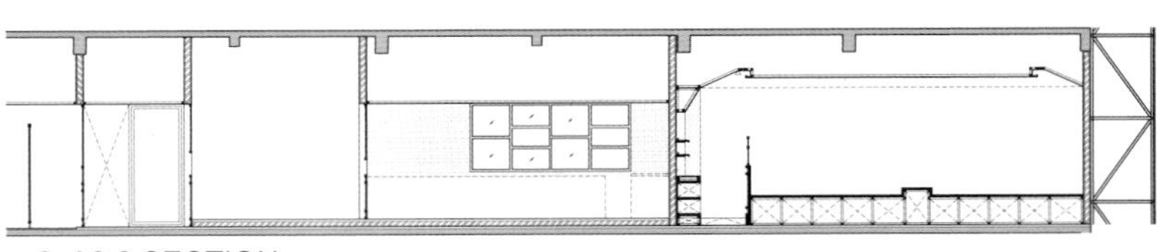

2 A3.2 SECTION

餐厅分为室内和室外两大区域。它还附带一个独特的空间，一间名为“The Class”的教室，专门教授那些对烘焙蛋糕、面包和点心感兴趣的孩子。室内区域包括开放式厨房、中心用餐区和位于后部的贵宾室。户外区域包括开放式酒吧和餐饮区。

走进餐厅，可以看到被天然石头覆盖的餐厅，几乎所有的墙壁都有浅蓝色的窗户，引人注目。长木条和水泥天花板连接着室内与室外这两个区域。人们会不知不觉地被地板材料的趣味性所吸引从而进入贵宾室。进入房间，随即被贵宾室里由绿色马赛克瓷砖和黄铜框构成的超长公共桌吸引，它最多可容纳 16 人，并配以水晶吊灯来渲染房间的气氛。放置在贵宾室中心带有黄铜框架的印花瓷砖成为整个房间的焦点。

木质天花板上有着坚实线条的艺术装饰风格门窗和以木材和天然石头为主的混凝土天花板将室内外区域分隔开来。以木头和天然石头为主的室外区域充满了蓝色元素，给人一种新鲜感和放松感。作为一张对北干巴鲁的美丽天气真实写照的画布，用五颜六色的家具作为色彩散点来呼应蓝天，像蓝天下的花朵一样。在室外用餐区，开放式的酒吧也在建筑周围设有公共酒吧桌，供给那些想要享受北干巴鲁白天或晚上美景的顾客。室外区域中间还有一个三角形天窗，可以增加光线，让空气流通。总之，La Fusion 是一个和合的地方。

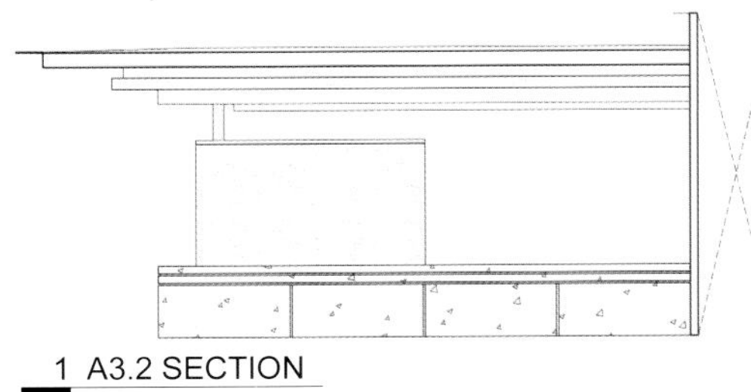

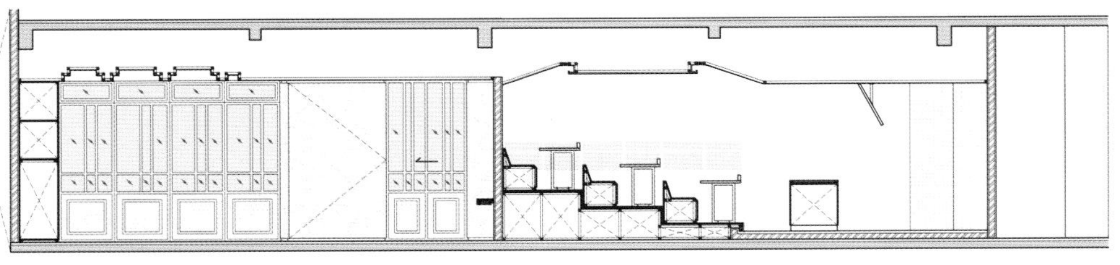

1 A3.2 SECTION

A REINTERPRETATION OF THAILAND STYLE

解语泰国风

Project Name | 项目名称
NAM THAI KITCHEN

Design Company | 设计公司
EINSTEIN & ASSOCIATES

Designer | 设计师
LEO EINSTEIN FRANCISCUS

Area | 项目面积
450 ㎡

印度尼西亚 雅加达
Jakarta
INDONESIA

Project Location

The concept of the design is the reinterpretation of a traditional house in Thailand, to combine contrasting aspects to complement each other. The name of the restaurant is "Nam", meaning "Water" in Thai; while the logo is a simplified shape of an orchid, a famous flower in Thailand that symbolizes timelessness and luxury. The design revolves around Thailand; whether it is the weather that is tropical, the atmosphere, the shapes, the color purple from the orchid that represent respect, power and loyalty or the style of the finishing materials that are mostly used in tropical style design.

该案例的设计理念是对泰国传统建筑的重新诠释，结合不同的方面相互补充。餐厅的名字为“Nam”，泰文的意思是“水”。它的商标是兰花的简化形状。兰花是泰国的著名花卉，象征着永恒和奢华。设计围绕泰国展开，无论是其气质、形状，还是颜色都象征着尊敬、权力和忠诚的热带兰花，装修材料的风格都采用了典型的热带风格设计。

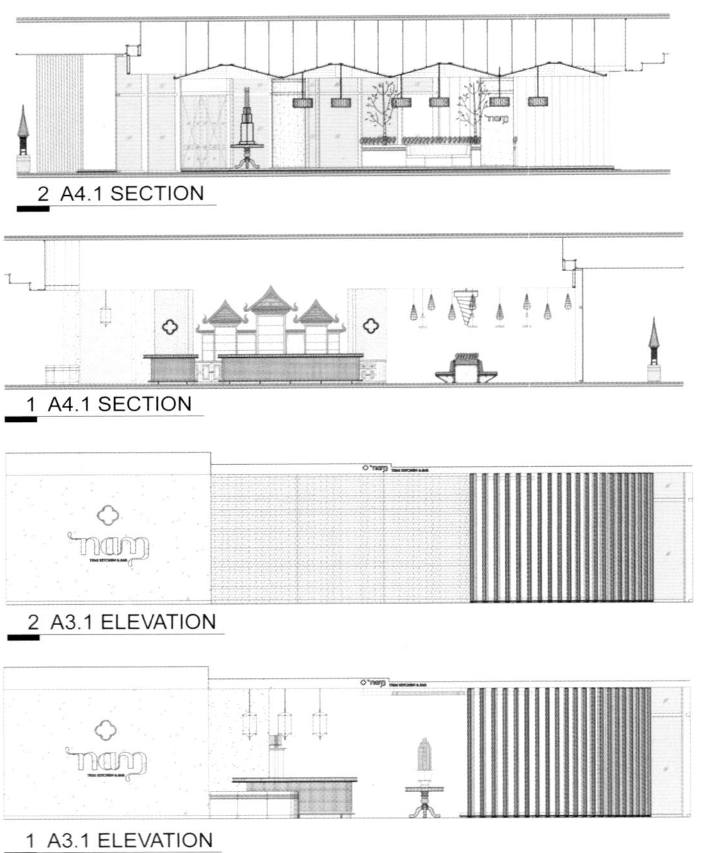

The whole restaurant acts as one big porch of a new interpretation of a Thai traditional house. Just as people are entering the space, whether it is the front or the back, they are welcomed by the lion statues and can instantly notice the bar, accented with shapes of Thai houses symbolizing the facade of a Thai traditional house. The other elements that people immediately notice after entering are the modern geometrical shape of the ceiling; mimicking the shape of an umbrella, protecting people below from the heat or rain, just like a porch. People can also feel the transition of the space, from the indoor to the outdoor at the back of the restaurant where there is a square with four benches at each side of the square. There is a stunning crystal chandelier resembling a waterfall over the center of the square. While the indoor area of the restaurant simulates the porch; the outdoor area simulates the garden and the transition area connects the two, combining it so that people can adjust immediately between the two.

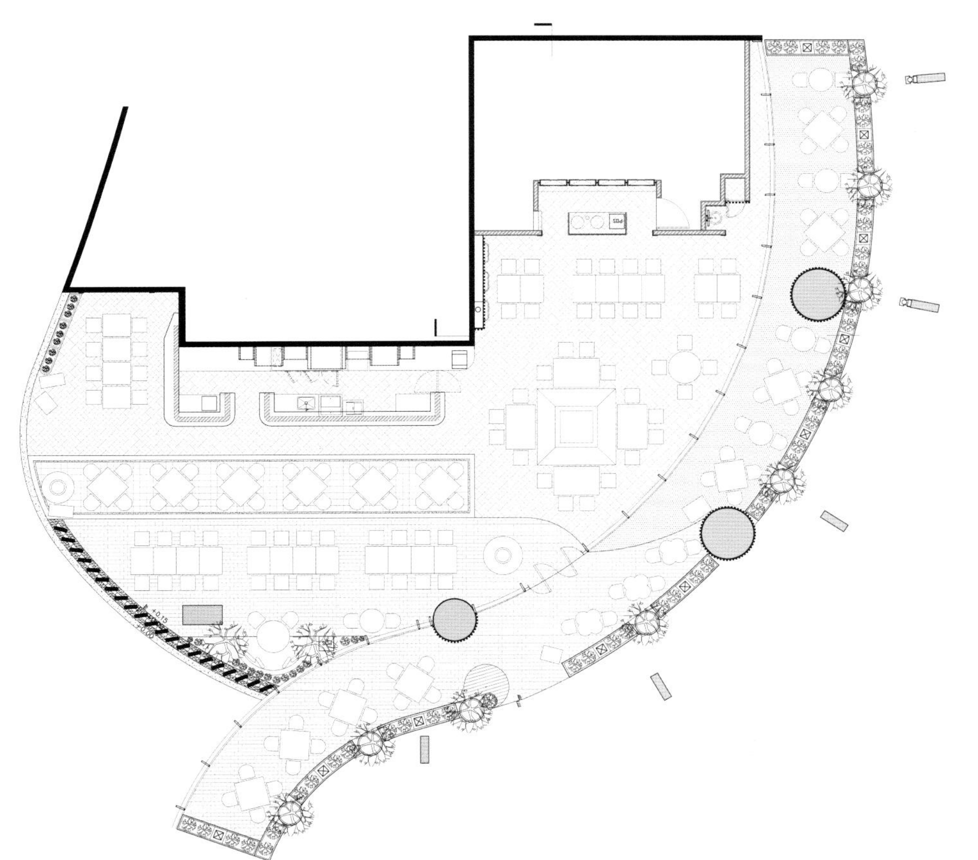

整个餐厅像是一个对泰国传统建筑大走廊的重新解读。不管人们是从正面还是背面进入餐厅，迎接他们的都是狮子的雕像，随即便可注意到象征着泰国传统建筑外观的酒吧。顾客们还会注意到现代几何形状的天花板，它是模仿雨伞的形状来保护顾客不受炎热和雨水的影响，就像门廊的作用一样。人们也可以很快感受到空间的转变，餐厅后面从室内到室外的过渡区是一个放着四个长凳的广场，那里有一个惊人的水晶吊灯，像瀑布一样垂坠在广场的上方。过渡区将像门廊一样的室内区域和像花园般的室外区域连接起来，人们可以很快适应。

The finishing materials used in the design are rich in color. Purple textured paint in the shape of Thai traditional roof is used to portray the elegance and calmness of the porch, terracotta, cement tiles, herringbone patterned basalt stone and rustic recycled wood are used all throughout the restaurant with a brass accent as the details.

Using the concept that the restaurant is a modern porch of a Thai traditional house, the designer uses a lot of contrasting details to invent a new style, to picture the modern while still embracing the traditional aspect of it all. The design explores the traditional aspect of Thailand then incorporating it into the design in a modern way. There are several contrasting elements in Nam, like the traditional house shape at the bar combined with mosaic and brass accents to not look too outdated, the folk style story in the form of murals, the custom outdoor light in the shape of traditional houses, the Thai's temple brass bells repurposed as hanging lights and the use of recycled wood to be more sensitive to the awareness of our today's environment. In conclusion, the designer hopes that this design can give a new interpretation of a Thai restaurant, not too modern that the beautiful element of Thailand disappear, and too traditional that it is only full with accessories. In this case, maybe it's best to combine the best of both aspects.

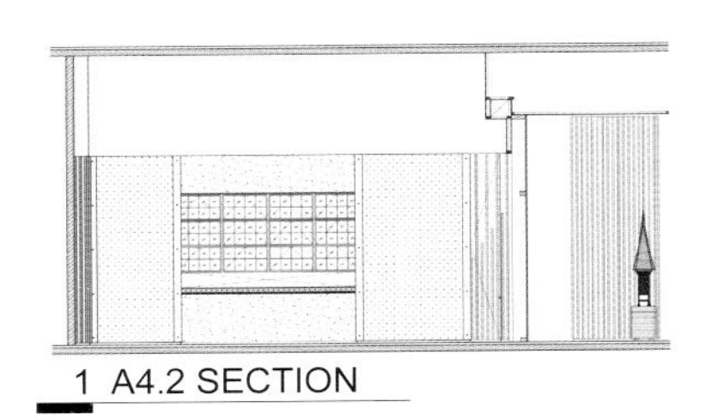

1 A4.2 SECTION

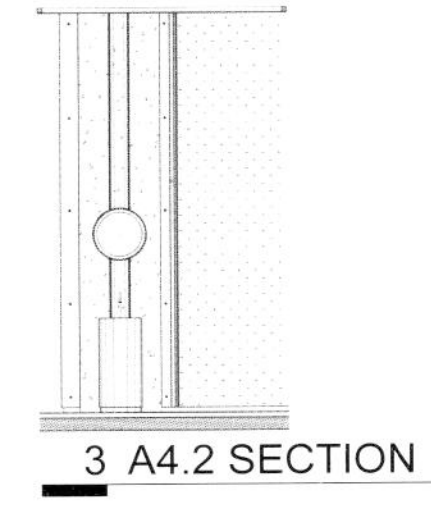

3 A4.2 SECTION

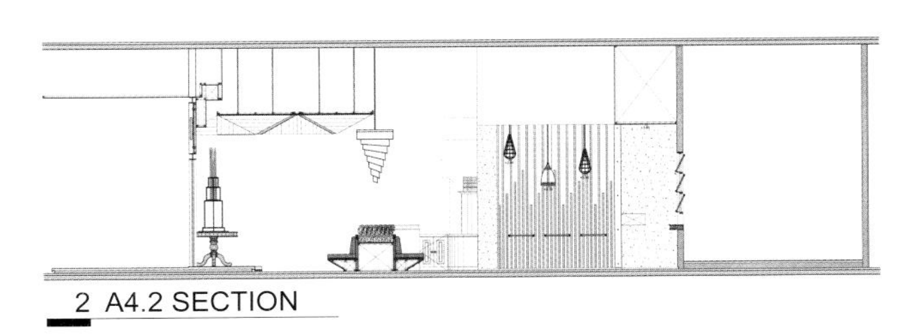

2 A4.2 SECTION

设计中使用的饰面材料色彩丰富。在泰国传统屋顶形状上，用紫色纹理涂料来描绘优雅沉稳的门廊，水泥、瓷砖、人字纹玄武岩石和质朴的可回收木材贯穿整个带有黄铜细节亮点的餐厅。

该案例结合餐厅作为一个泰国传统住宅现代门廊的理念，设计师使用了大量的对比细节来营造一种新的风格，在描绘现代的同时仍然拥抱传统。该设计探索了泰国的传统面貌，并以现代的方式融入其中。Nam有几个极具对比性的元素，比如酒吧里传统房子的形状搭配镶嵌黄铜亮点使其显得不太过时，用壁画形式来描绘民间故事，传统房子形状下定制的户外灯，泰国寺庙铜铃挂彩灯和可循环利用的木材都体现了我们正逐步提高当今的环境意识。总之，设计师希望这种设计可以带给泰国餐馆一个新的诠释，不至于太过于现代导致泰国本土元素的消失，又不会太传统到只充满了配饰，在这个案例中，都得到最好的诠释。

THE ELEGANT AND OLD FASHIONED CLUB RESTAURANT

典雅老派的俱乐部餐厅

Project Name | 项目名称
Studio Hermes

Design Company | 设计公司
Corvin Cristian

Designer | 设计师
Corvin Cristian

Area | 项目面积
200 ㎡

Photographer | 摄影师
Cosmin Dragomir

Corvin Cristian designed the interiors of Studio Hermes, a club and restaurant located in Bucharest, Romania, that host a variety of shows from cabaret to live bands. The design responds to acoustic requirements hence the look recalling an audition hall in the 1960s. Other design elements, while contemporary, follow the same mid-century modern line. The onyx bar, velvet sofas, brass cymbals and walnut wainscot as a counterpoint to the bare concrete and exposed piping. Studio Hermes is located in Bucharest's historical centre and takes its name from an old movie theater that was formerly located at the same address.

Studio Hermes餐厅及酒吧由设计师Corvin Cristian设计，位于罗马尼亚的首都布加勒斯特。在这里有许多的表演，既包括歌舞，也包括现场乐队。因此出于音效的考虑，其设计有点像60年代的试唱厅。其他设计元素，虽然是现代设计风格，但是不乏中世纪设计元素。黑玛瑙色的酒吧、天鹅绒沙发、铜钹、胡桃色的护壁板与裸露的混凝土和管道形成了鲜明的对比。Studio Hermes餐厅坐落在布加勒斯特的历史中心，旧址是一家老电影院，也得名于此。

This restaurant contains three floors with spiral stairs connecting all floors. There are performance areas, seating areas and bar areas on the first floor; DJ console and seating areas on the second floor, and the seating areas and bar areas on the third floor. In the middle of the dining room, there is a double-height oval atrium, and the walnut planks are erected along the circumference, dividing boundary and forming handrails on the second floor and the ceiling decorations on the first floor. The custom-made copper chandeliers hang down from the atrium, which is elegant and charming. The allocation of copper chandeliers and walnut renders a warm and relaxed atmosphere. Here, you can enjoy the delicacy, or taste a good wine, or enjoy a performance.

整个餐厅一共有三层，一个旋转楼梯联通上下，一层有表演区、座位区和酒吧区，二层有DJ控制台和座位区，三层有座位区和酒吧区。餐厅中部有一个双层挑高的椭圆中庭，胡桃木板沿四周架设，划分边界并装饰二层扶手和一层的天花板。定制铜吊灯从中庭垂落下来，优雅迷人。在铜吊灯和胡桃木的搭配下，整个空间给人一种温暖放松的气氛。在这里，你可以享受美食，或品一杯好酒，亦或是欣赏一场演出。

THE GOLDEN WARMTH,THE ROMANTIC COBALT

温暖流金 浪漫海色

Project Name | 项目名称
多伦多海鲜自助餐厅（万象城店）

Design Company | 设计公司
上瑞元筑设计有限公司

Area | 项目面积
920 ㎡

Main Materials | 主要材料
理石、金属帘、木地板砖、花砖、喷塑铁板、瓦楞玻璃等

The project is situated a new shopping business district of new consumption experience patterns of the beautiful and pleasant Binhu District, enjoying competitive geographical advantages. The project is characterized by the flexibility of field application in terms of plane design, organic link and the perfect combination of platform area and seating area. It divides detailed fields in combination with flooring, hanging mental chains and vertical screens.

Metropolitan light and luxurious style become the required internal deposit. It attempts to capture the new visual experience in the metropolitan catering environment, extracts mental components and elements in space; converts the leather, mirror face, fabric into a delicate artistic connection. The combination of tasteful mental structural accessories and the mental curtain in a skilful manner is converted into a visual presentation. Among the elements of space, units, colors and lights, the warm and elegant sense is highlighted, offering more ease and comfortable atmosphere.

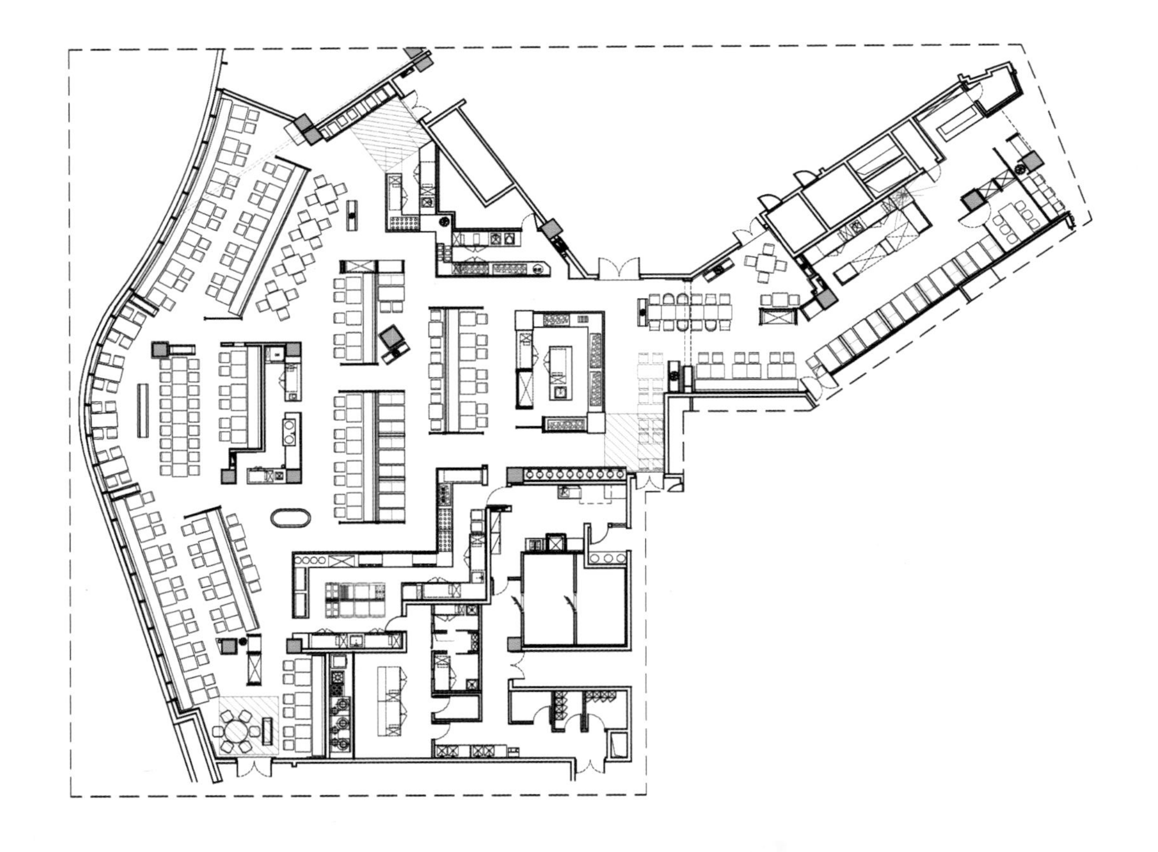

本案地处景致宜人的滨湖区，一处拥有体验式消费的购物商区内，坐拥优良的地理优势。全案在平面配置上考量场地运用之灵活性，岛台区域与座位区有机结合，完美衔接，动线布置的串联结合地坪、金属挂链和立屏进行细部场域划界。

都会轻奢风格是本案的内在底蕴，试图捕捉都市就餐环境的新感观体验，在空间中提取金属构件元素，将皮革、镜面、布艺等进行精美的艺术拼接。造型考究的金属构架饰品与金属挂链巧妙串联，两者结合而转化为视觉引线，令它们穿引在空间、装置、色彩、光影的序列之中，烘托出就餐环境温暖又典雅的氛围，使空间整体气氛更加轻松自在。

With careful observation and individual experience as the point of start, the design integrates both western and eastern history and intent of modern cities. Visual modern and luxurious expression make the project embraced in a modern fashion atmosphere, making guests indulge in the metro life sentiments while enjoying delicious dishes.

设计以细腻观察与个人经验为出发点，复合了东西方往昔与现代都市的意向，轻奢摩登的视觉语法令全案浸润在时尚的都市气息中，使宾客在享受美景佳肴之际，沉醉于大都会的生活情调之内。

ENJOYING A CHINESE STYLE RHYME, RELEASING THE INTERNATIONAL STYLE

品中式韵 释国际范

Project Name | 项目名称
HI 辣火锅

Design Company | 设计公司
寸 DESIGN

Designer | 设计师
崔树

Area | 项目面积
300 ㎡

Taking the traditional culture of Chongqing as the keynote, the designer references the image elements of the Millennium town Ciqikou, the epitome of Chongqing. He integrates the characters of architecture, life and diet culture, combines with modern techniques in the fusion of interesting and the appreciation of Chinese culture to endow the cultural connotation to the public, add social levels and create a unique and original environment.

The spicy red is an exhibition color in the space, and the background color and beautiful color are harmony and unity. The scattered red lights map down with the shadowy halo with wooden grid window. The Oriental elements are everywhere, and a reasonable collocation of materials and colors draws an antique style to make hotpot have a historical precipitation and propagation.

Hi辣
Hi辣
Hi辣

Hi辣
Hi辣
Hi辣
Hi辣

Hi辣
Hi辣
Hi辣

Hi辣
Hi辣

Hi辣
Hi辣

Hi辣

Hi辣
Hi辣

空间以重庆的传统文化为基调，借鉴有重庆缩影的千年古镇磁器口的形象元素，融合建筑、生活、饮食文化特色，同时结合现代的手法，使现代的趣味性与中国文化的欣赏性相融合，为大众化消费赋予文化内涵，增添空间文化层次，打造独特新颖的环境。

辣的红色是空间的展示颜色，背景色和点缀色和谐统一。错落不齐的红灯红，映射下来的光晕影影绰绰，配合栅格木窗，东方元素弥漫在各处；材质、色彩搭配合理，绘出古色古香的风情，让火锅亦有历史的沉淀与传播。

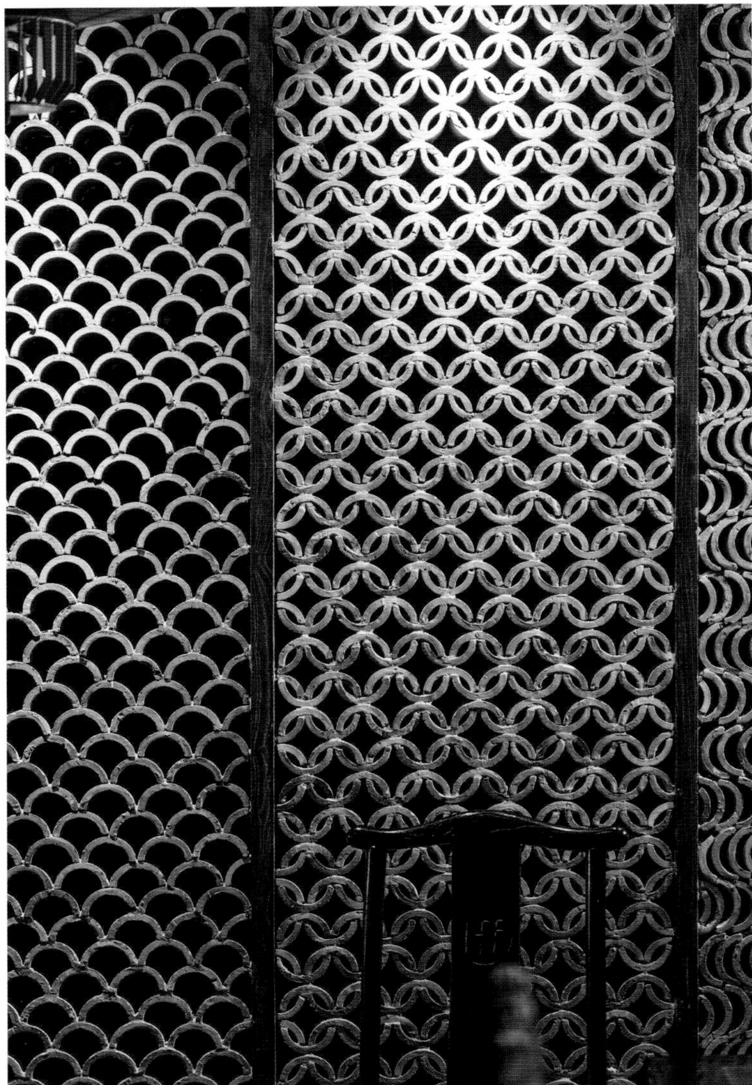

The symmetrical pattern is elegant, and the long and narrow corridor brings the visual perception of the scene changed at every step. Besides, the simple and elegant structure and the full-tasting and mature color make the space large yet not empty, thick yet not heavy. It's stylish but not depressing, which not only restores the purest beauty, but also deduces an international style.

对称式格局显得高雅，狭长的走廊带来移步换景的视觉观感。此外，设计造型简朴优美，色彩浓重而成熟，使得空间大而不空、厚而不重，有格调的同时又不显得压抑，不仅还原最朴实的美，还演绎出国际范儿。

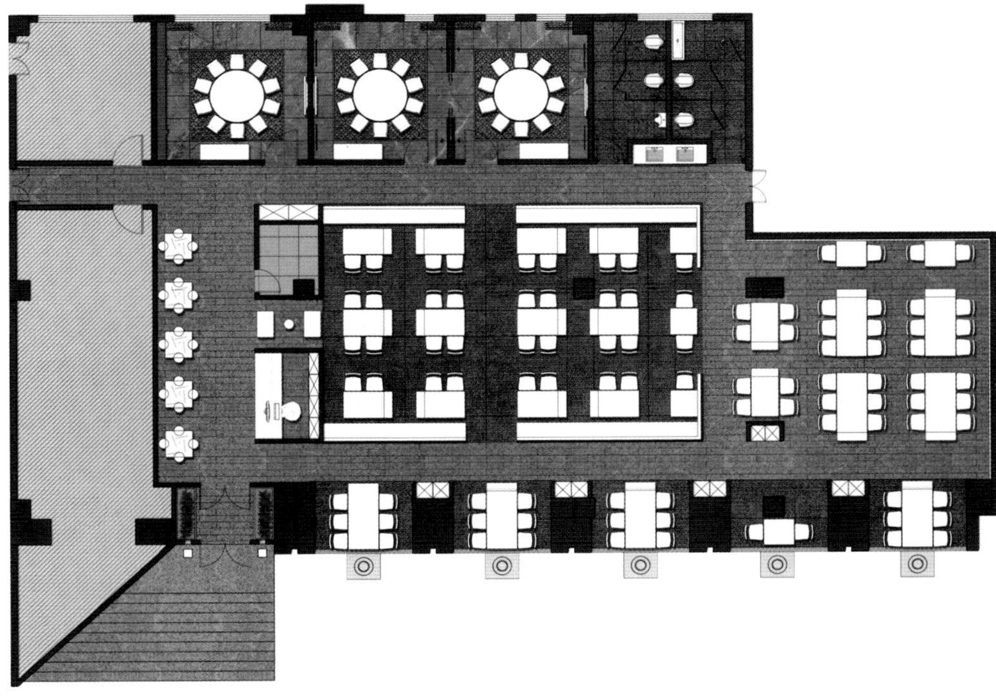

THE POWER OF DEVOTED DESIGN

匠心温度

Project Name | 项目名称
葫芦岛食屋私人餐厅

Design Company | 设计公司
纬图设计有限公司

Designer | 设计师
赵睿

Area | 项目面积
2101 ㎡

Main Materials | 主要材料
片岩石、松木、铁艺等

Photographer | 摄影师
杨戈

The architectural style of "Food House" is a typical and retro architectural style in the 70s and 80s. The location of this case is relatively advantageous with the broad vision and the vast seascape outside the window. The owner's main expectations are the remodification and extraction on the appearance of the building. A building should combine with the new functional requirements, echo with the surrounding environment and form a unified temperament from indoor to outdoor, so that can integrate into the situation better.

First of all, the designer begins to design from the generatrix of the building inside and outside, and believes that design is to make people use the product itself well, so that the analysis of the generatrix of the building inside and outside is the most crucial problem to solve. The original entrance steps straight to the road will be cleared and placed on the west side of the building. Up to the stairs, the stone lions symmetrically distributed on both sides, add a lot of interest and a sense of ceremony for the entrance. In the process of building the generatrix of outdoor terrace, the designer avoids the existing plant landscape as far as possible. Because of this, the form of making plane outline has one more "natural" sense of belonging of place. After the remodification, the designer uses the gray-based building appearance and the trapezoidal trend of the building volume to integrate the building and the surrounding landscape into a unit and complement each other.

"Food House" is positioned as a private club for owners gathering relatives and friends to use for the party and not for commercial purposes. So there is no need to make interior decoration cater to everyone's taste deliberately, which also provides a relatively free space for the "process" creation and the release of his emotions to the designer. Of course, based on the solid practical abilities of the designer, everything seems to be in control.

Straw, a very daily and simplicity plant, with its single form is not mighty but a little weak. While when it forms a continuous replication and array distribution, it will show an unbalanced sense of power on the whole. This state is like daily labor, and the process seems to be repeated and uninteresting. With days and months multiplying, some specific experience and wisdom thus exist. Why not the design work? In contrast to the specific forms of the project shows finally, the designer pays more attention to the generation of logic between the design process and the form breeding from the process. This is the fun of design. The designer tries to transform the spirit of "straw" into a space construction language integrating into the narrative of "Food House", and then appears the "straw" device in the entrance hall, as well as the continuous irregular wood texture in the walls of every space node and ceiling. Based on the three-dimensional spread of such device node design and the same form, allocating with the designer's improvisation, the designer makes the boundary of space creation extended to a certain extent and gains more unknown and exploratory of artistic creation.

“食屋”项目建筑样式为典型七八十年代复古建筑风格，项目所在地理位置相对优越，视野开阔，窗外直面无边海景。该项目业主委托的主要期许是对建筑外观重新进行修整和提炼，建筑应结合新的功能要求，对周边环境有所回应，做到室内外形成统一的气质，让建筑更好地融入到环境中。

设计者首先从建筑内外动线着手展开设计，坚信设计是为了让人更好地使用产品本身，所以对建筑内外的动线梳理便是首要解决的问题。原本直冲大马路的入口大台阶将被清除并且设置在建筑的西侧，拾级而上，两侧对称分布的石狮为入口增添了不少趣味和仪式感。在户外阳台动线的增建过程中设计者尽量避开场地现有的植物景观，正因如此，建筑平面轮廓的形态多了一份“自然而然”的场地归属感。经重新改造后以灰白色系为主的建筑外观，以及建筑体量的梯形趋势展开，使得建筑与周边景观融为一体，相得益彰。

安全出口

“食屋”定位为私人会所，供主人和亲友在此聚会使用，并不考虑对外商业用途。所以无需刻意去让室内装饰迎合众人口味，这也为设计者提供了一个相对自由的空间来进行“过程式”的创作和自我情绪彻底的释放。当然，基于设计者足够扎实的实践功力，一切却似乎尽在掌控之中。

稻草，一种极为日常和朴素的植物，它单个形态并不强势反而显得些许瘦弱，但当它形成一片并不断复制阵列分布时，在整体上便会呈现出一种非均衡的力量感。这种状态好比日常劳作，过程看似重复和无趣，日积月累，某种具体经验和智慧由此孕生。设计工作又何尝不是这样呢？对比于最终项目呈现出的具体形式而言，设计者更注重设计的过程性以及籍由过程所滋生的形式之间逻辑性的生成，往往设计的乐趣就在于此。设计者尝试把“稻草”具备的基本精神置换成一种空间构筑语言融入到“食屋”整个空间的叙事中去，进而出现了入口前厅的“稻草”装置，以及在每个空间节点的墙身和天花上延续的不规则木条肌理。设计师希望基于这样的装置节点设计及同种形式三维式地铺开，再加上设计者的现场即兴创作成分，使得空间创作的边界得到了一定程度上的延伸，多了些艺术创作的未知性和探索性。

When space is formed by the "straw" gray background, the other smart nodes rely on some rich texture objects as a highlight to render the space atmosphere. For example, a group of exercising of "monk" sculptures in the window, with the reflection of light in the side window, the original head in blurry shape presents an outstanding and abundant expression. Despite the rational creation factors, the designer is more willing to believe that this scene is a coincidence at this time, this place, and this scene.

The designer thinks that the concept of design must be existing before design. The concept seems to be several fragile thoughts, and under the weaving of logical consciousness, they establish an inner connection, work with each other and form a complete harmonious state. In the traditional Chinese concept, it is significant that people is faith in and worship the "craftsmanship". "Craftsmanship" does not mean a flavor of "craftsman" or the traditional technical levels. In fact, it advocates a kind of spirit which is the process to explore and dig out the value of daily and accessible items, and finally, they produce a new structural relationship. The designer tries to touch this state, and the embodiment of "Food House" is the integration of specific items and space. The glass lamp large contrasting with wood color, the deadwood picked up from the roadside, and shells after eating and others are recreated to form a three-dimensional relief wall. The white transparent plates and goblets on the table, the small gourd buying from the market and so on, all of these present a breathable overall sense. Obviously, what matters is not the single object itself, but the "special concept" that is deeply embedded and deepens in the designer's brain.

Finally, this "Food House" project is emphasized a design view of the designer: the final result of every design project is only a concentration and phased embodiment of the designer's real status. The time is going, and the concept is growing. In the process of practice, only when the designer keeps the open-minded thinking and continuously speculate himself, can he have the opportunity to realize the works full of temperature and vitality.

当空间被“稻草”所形成的灰色调背景完成后，其他灵动的空间节点便是靠一些质感丰富的物件摆设来担当角色，进一步烘托空间气氛。例如窗台一组晨练中的“和尚”雕塑群，在侧窗一缕光线的作用下，原本形态模糊的头部呈现出的表情格外出彩和丰富。抛开理性创作因素不说，设计者更愿意相信这种景象是一种巧合，巧合在于：此时，此地，而后此景。

设计者认为，设计之前，必须观念在先，观念是看似零碎的若干想法，在人意识逻辑的编织下，它们建立起某种内在的关联性，彼此合作，共同发力来形成一个完整的和谐状态。中国传统观念中人们对于“手艺”的信仰和推崇甚为显著。“手艺”并不意味着带有“匠气”味或是指向因循守旧的某类技术层面。实际上，它宣扬一种精神，那就是对日常的、垂手可得的物品的价值的探索和挖掘过程，最终让它们产生一种新的结构关系。设计者试图触及这种状态，在“食屋”空间中的具体体现便是极具差异性的物件与空间的共融：与木色反差较大的玻璃工艺灯，路边捡来的枯枝和食用后的贝壳等物件经现场再创作形成的立体浮雕墙，包括桌上的白色碟子和透明高脚杯，市场上淘来的小葫芦等等空间里的一切物件呈现出一种透气的整体感。很显然，重要的不是单个物件本身，而是深植于设计者脑中并且不断深化的“空间观念”。

最后借“食屋”项目来强调设计者的一个设计观点：每个设计项目最终所呈现出的结果只是设计师当时真实状态的一个浓缩和阶段性体现，时间在推进，观念也在生长。设计师只有在实践的过程中保持开放的思维状态并且不断地进行自我思辨的情况下，保持诚恳，才有机会实现富有温度和生命力的作品。

FREE, WILD AND PUNK STYLE

洒脱粗犷轻朋克

中国 昆明

Kunming CHINA

Project Location

Project Name | 项目名称
醉桃 · 餐 · 音乐酒馆

Design Company | 设计公司
昆明鱼骨设计事务所

Designers | 设计师
纳杰、吴浪

Area | 项目面积
530 ㎡

Main Materials | 主要材料
毛石、艺术涂料、原木、金属等

Photographer | 摄影师
许峻玮

TAO Bar is located in the most beautiful Guilonghu Park of Kunming which its external environment is fulfilled with flowers. The interior design uses a retro style, adopting the cement elements with iron accessories and combing with exposed ashlar to represent a rough texture. Locomotive mode with bottle caps gives the restaurant a free temperature, whilst the graffiti on the wall enhances the art aesthetic of the whole space. Designers add the rust red color into the dark gray cement space to show a bright yet vivid atmosphere.

醉桃 · 餐 · 音乐酒馆坐落于昆明最美的龟龙湖公园，外环境花团锦簇，繁花似锦。整个室内空间设计采用复古的风格，运用水泥元素搭配铁艺饰品，结合墙面裸露的毛石，给人一种粗犷豪迈的质感。贴满瓶盖的机车模型赋予了餐吧一种洒脱的气质，而墙面的手绘则衬托了整个空间的艺术美感。设计师在深灰色的水泥材质空间中加入了复古的铁锈红，让整个空间既轮廓清晰又不失活泼的气氛。

This is a two-storey restaurant bar. The first floor consists of the individual seats and the rectangular table area. Besides, opposite to the stage, there is an island bar for people who drink only sitting on the seats under the stage to enjoy the performance. The second floor mainly consists of the rectangular table area and two private rooms. Besides, there is a large terrace which guests can sit along the lake, enjoying the pleasant dining environment. For its restaurant-bar operation model, the overall lighting system is controlled intelligently. This design has a perfect combination of indoor and outdoor, and gives guests a touch of unique feeling.

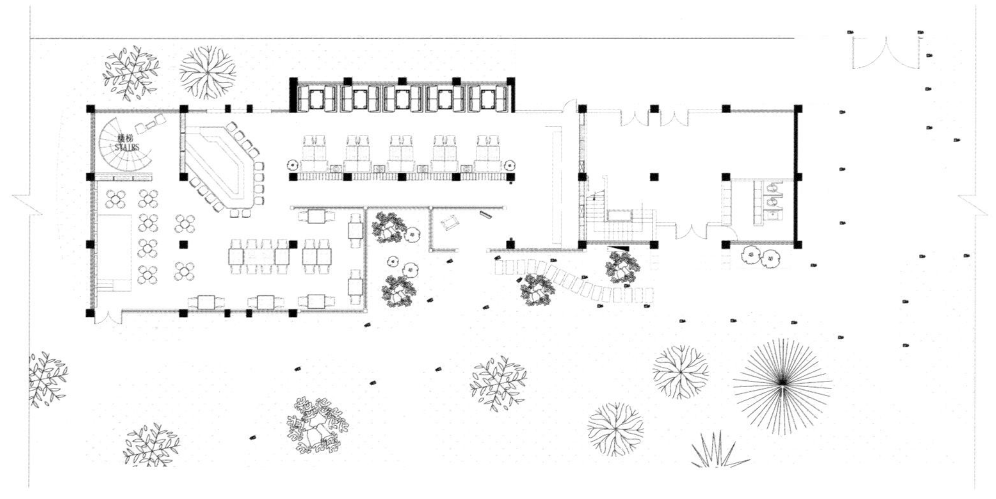

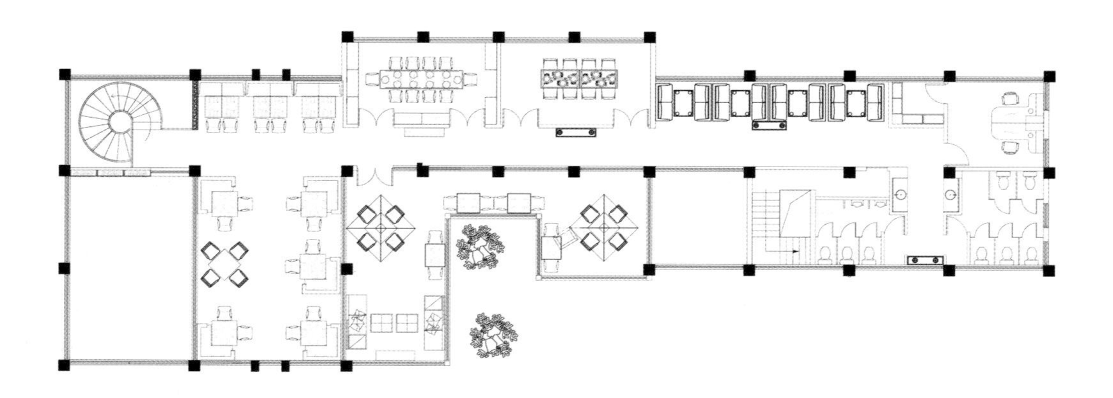

餐吧为两层，一楼由散座区、卡座区组成，另外有一个岛式吧台与舞台相对而立，只为喝酒而来的客人们可以坐在舞台下方的散座区欣赏驻场的表演。二楼主要由卡座区和两个大包间组成，另有一个大露台，客人在这临湖而坐，就餐环境更是赏心悦目。因为是餐带酒馆的运营模式，因此设计师对餐厅的整体灯光运用做了智能化调控。设计整体让室内外空间完美结合，给客户一番别样感受。

THE RABBIT HOLE IN THE MOON

月球上的兔子洞

中国 澳门
Macau
CHINA

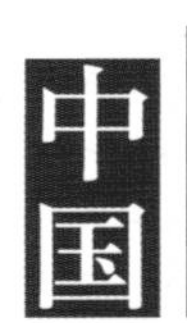

Project Location

Project Name | 项目名称
Le Lapin

Design Company | 设计公司
威尔逊室内建筑设计公司（纽约）

Designer | 设计师
Dan Kwan

Associate Designer | 助理设计师
Christina Wang

Graphic Designer | 平面设计师
Holland Hames

Photographer | 摄影师
Benoit Florencon

In many times, what we shared is not only just a good meal but also a matching space and atmosphere. The first impression of Le Lapin must be "playful". Guests are greeted by two rabbit guards at the door before marching in the restaurant. Guests might find themselves in the mysterious rabbit hole, a hole that is on the moon.

The entire concept started with our client Carson, who was born in the year of the rabbit. Carson wanted to open a modern French restaurant in the Macau Science Center, a distinctive conical-shaped building designed by famed architect I.M. Pei.

很多时候，我们分享的不仅仅是食物的美味，还有与之相配的空间和氛围。调皮，这一定是所有来宾在看到Le Lapin餐厅第一眼时的共同印象。他们会忍不住眉梢眼角的笑意，伸手推开由两只兔子把守的大门，迈步走进餐厅。他们一定会以为自己来到了兔子洞，还是建在月亮上的兔子洞。

整个设计理念从兔年出生的业主Carson说起，Carson一直有要在澳门科学馆内开一间现代法式餐厅的想法。澳门科学馆由享誉全球的著名建筑师贝聿铭设计，圆锥形的建筑独具特色。

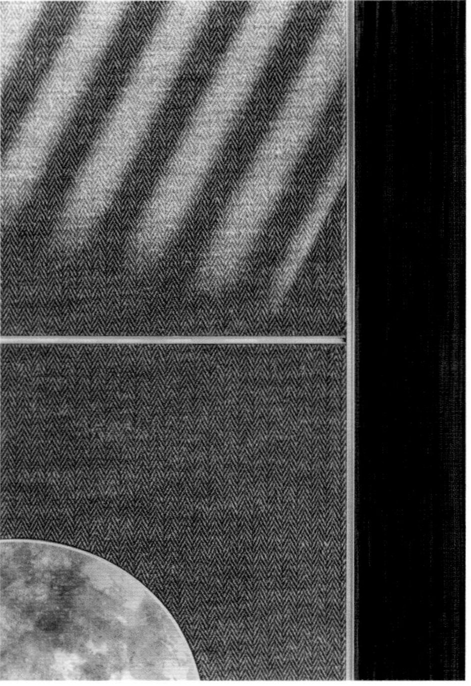

Influenced by the restaurant's circular shape, the owner's Chinese zodiac, and Macau's status as an intersection between East and West. The design team arrived at two stories: a story of traveling to the moon written by famous writer Jules Verne, and the beloved Chinese fairy tale of the Jade rabbit in the moon.

In Chinese folklore, the rabbit is the companion of the moon goddess. This relationship fueled every element of the design, from the entryway to the private dining rooms and restrooms. Not only would guests leave talking about the savory French cuisine, but they would also start to leave talking about the design symbol and playful design narrative.

Then came Mr. Rabbit or "Monsieur Rabbit", the protagonist, the mascot, the inspiration for this tale. Mr. Rabbit is a French sophisticate, a wine aficionado, and a bit of a cad. His image as a symbolism was represented by turn-of-the-century French lithographs. His French mantra "jenous se quais" is found everywhere in the entire venue, especially in luxury, yet quirky finishes and furnishings.

MGM

Like the moon, Le Lapin is laid out in a perfect circle. Guests exit the elevator and arrive in a French art gallery setting with a piece of eye-catching hanging artwork: flying candle crystal chandelier.

Then, they will see a spectacular "wine wall". This wall is 16 meters high and inventively showcases Carson's wine collection worth more than 2 million dollars. It is a marvel to behold! It's three levels of catwalks and various LED/fiber optic starlight elements that play into the lunar design motif like stars in the sky. It is the heart and soul of the entire venue.

Guests walk along the wine wall and into the bar-lounge area, which exudes French sexiness. The tables are tall, the floor is dark marble, while the bar itself is designed to be a floating crescent moon of backlit onyx. The design team decided to use the crystal whiskey bottles in mirror boxes of varying sizes behind the bar, so that they wouldn't obstruct the views of the Macao skyline. The main dining parlor, separated by airy gold mesh panels, is an elegant, tailored space with a neutral color palette that seats 50 guests.

The wine tasting room is a double-height space with a floor-to-ceiling collection of books, thematic artwork, accessories and wine refrigerators. The walls also feature trap doors and "peek-a-boo" windows that look into the private dining rooms, adding a touch of playfulness.

The carpet in both dining rooms is a piece of conversational artwork inspired by poetry. The carpet in the large private dining room is based on Jules Verne's De la Terre à la Lune (*From the Earth to the Moon*) and features a collection of custom-designed hot air balloons, while the carpet in the small private dining room features Chinese poet Su Shi's *The Mid-Autumn Festival*. In fact, the poem is angled in such a way that guides one's eye out over the Macao skyline, obliging the reader to stop and reminisce. Again, the intent is to create a truly interactive dining experience that leaves the patron with enduring memories, not just a good meal.

受餐厅圆形的形状、业主的生肖和澳门本身特有的中西结合的元素启发，设计团队想到了两个故事：由著名法国作家儒勒·凡尔纳所作，讲述到月球旅行的小说故事，以及中国神话传说中有关玉兔的故事。

中国民间传说兔子与嫦娥为伴，两者之间微妙的关系被融汇到了各个设计元素中，从入口通道到包房甚至洗手间都有所体现。客人们在离店时不仅会提及可口的法式美食，那独特的设计符号与有趣的设计故事更加令人回味无穷。

接着出现了一个主角，一个吉祥物，整个故事的灵感来源——兔子先生（或“兔子绅士”）。兔子先生是一名法国绅士，对酒的喜爱可谓是到了狂热的程度，但性格中又带点痞，优雅中又带有一些顽皮。以他的形象作为标志，用二十世纪初法国平版印刷将其呈现。整个场所内随处可见他的法语口头禅“je nous se quais”（我们在码头）特别是在一些奢华却又稀奇古怪的饰面和家具上。

就像月亮一样，Le Lapin的布局也是按照一个正圆展开的。一出电梯感受到的是一种法式艺术长廊式的基调，正中一个飞翔的蜡烛水晶吊灯艺术品特别吸引眼球。

紧接着看到的是一面壮观的“酒墙”。这面墙有16m高，用独特的方式展示了Carson的收藏，这些酒的价值超过两百万美元，光是目睹就已经感到惊叹。三层高的狭窄过道和各式LED/光纤星光元素加入到整个月亮设计主题中，就像天上的星星一样闪烁，这是整个场所的灵魂所在。

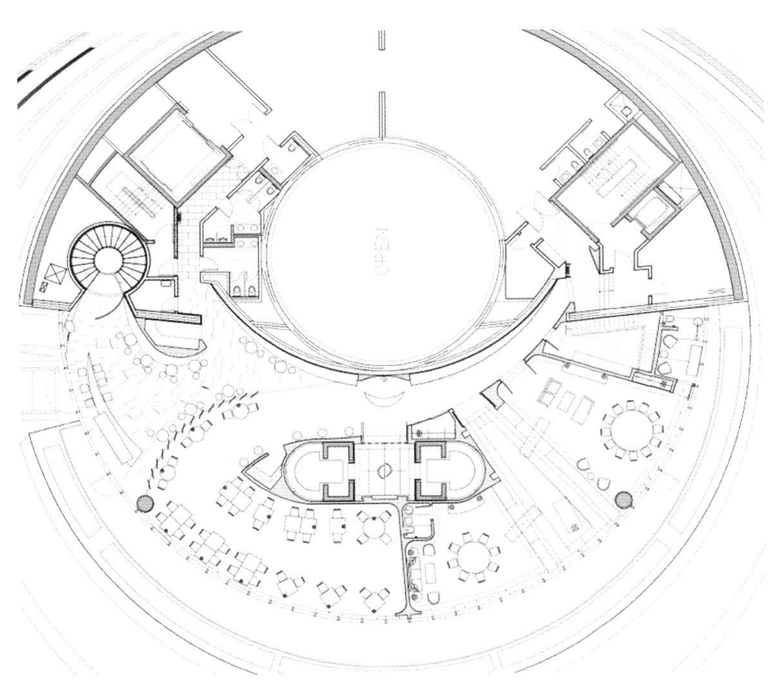

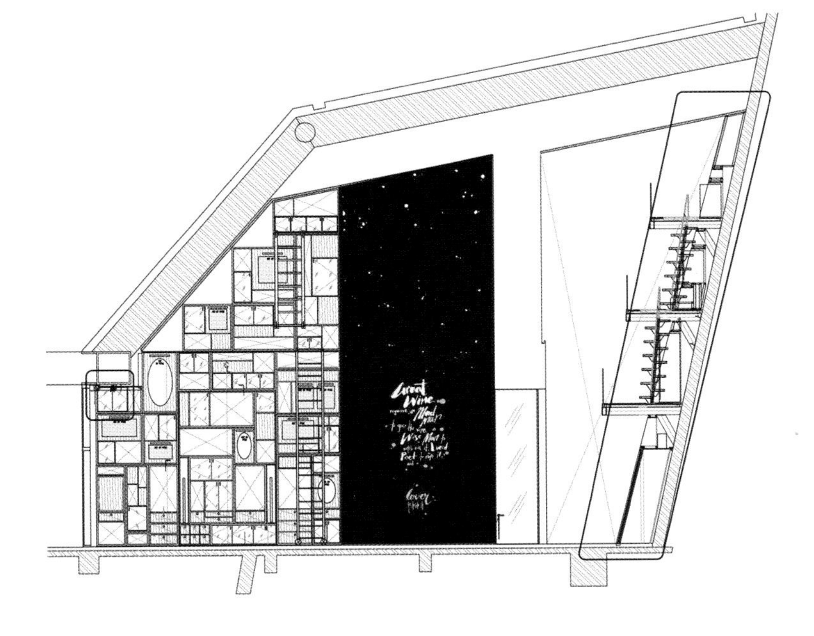

沿着酒墙一路走，便到了休闲酒吧区，整个区域自然而然散发出一种法式的性感。桌子相对较高，地面则使用了深色大理石。吧台本身使用背部打光的缟玛瑙，形似漂浮着的新月。设计团队选择使用大小不一的镜面方框来放置水晶的威士忌酒瓶，作为吧台的背景，既美观，又不阻碍人们观赏澳门的视线。主要用餐区以轻快的金色发光薄纱分隔，是一个典雅、精致的空间，以中性色调为主，可容纳50位客人用餐。

品酒室是一个双层挑高的空间，从地板到天花通高的展示墙面上，收藏着各种书籍、艺术品、配饰与红酒冷藏柜。墙面还带有活动暗门与“躲猫猫”似的小窗，可以瞥见包房内的情景，为整个空间加入了一丝趣味。

两个包房的地毯都可谓是一种艺术品，一种以诗歌艺术为灵感而生的艺术品。大包房地毯使用的是儒勒·凡尔纳的De la Terre à la Lune（《从地球到月球》），伴有特别定制的热气球，小包房地毯使用的是中国诗人苏轼的《水调歌头 明月几时有》。其实，朗读诗歌时所需的角度恰好在不知不觉中将人的目光领到澳门天际线上，非常体贴地让正在朗读的人不知不觉地停下回忆。设计意图是希望能营造一种真正带有互动的用餐体验，让人们在一顿美食过后，仍对整个空间流连忘返。

HONG MAISON RESTAURANT

民国红公馆

Project Location

Project Name | 项目名称
南京名谷设计机构

Designer | 设计师
潘冉

Area | 项目面积
1700 ㎡

Main Materials | 主要材料
木材、灰泥、水磨石、大理石等

Photographer | 摄影师
李国民

Continuity

In early 2016, the guest paid his visit in Jaco's office, hopefully, to create a flagship restaurant in the south of Nanjing City. The unforgettable charm sparkling in past Republican China essence was anchored in the space, called Hong Maison Restaurant. The place appears appropriate, featuring a sophisticated newly-built bridge joining two pavilions separately located on the sides of north and south. Water flows beneath the pavilions and splits into several brands. Behind them, there you can find pines and cypresses, Chinese bananas, black bamboos, peach blossoms, willows, etc.

【续】

丙申年初，吴先生来访，欲造店城南为旗舰，取民国风韵谓之精神，名曰民国红公馆。选址停当，有南北二楼架桥相连，流水其间亭院各处，后置松柏、芭蕉、紫竹、桃花、杨柳等。

红公馆
紅公館中餐廳

Field

Passing through the memorial gateway of the historic avenue, visitors arrive at the Jianzi Lane in the east after a thirty-meter walk. Ascending the stairs to reach the main door, there apparently standing a 1.67m wide door open to the interior. The entrance-hall with a semi-enveloped screen welcomes you to enter and appreciate the eclectic design with a mixture of affection and style, while incorporating the signature of the community. The living room catches large surprised eyes after making a right turn through the entrance hall. Bar and reception were arranged on the west and east side and set to provide visitors hospitality service. A free-standing island library centralizes and partitions the space to organize internal flow. Bar in the west side features exclusively made relief sculpture named 'Tale of Nanjing' from the local artists embedded in the background, characterizing scenes such as exorcising of evil spirits, avenues and lanes, rosy cloud and beautiful scenery of Qinhuai District. Etc. It starts to emerge the story about the past with these. Above the sculpture, a board-carving purple paulownia plaque written in Chinese calligraphy "红公馆" (Hong Maison Restaurant) on the wall was created by calligrapher Le Quan with a great reputation as 'contemporary calligraphist of China'. An old-fashioned phone made of lodestone stands on the bar, echoing the hung chandeliers above it. Details of old Encyclopedia, bird cage, terrine pot and candle holders arouse people to recall the most important parts of the history; perhaps, because of the customer's birth origin of the literary family. The reception area in the west anchors the fireplace with an old-time painting featuring the past president's office hung on the wall, consistent with the sculpture in the opposite. A colorful Chinese drum-shaped stool and jacquard-weaving carpet appear delicately pretty and elegant.

【场】

经老门东牌坊入剪子巷东30m处，由北拾阶而上，见五尺宽铜门向内，迎面玄关，屏风半掩，于转折处入公馆客厅，沿东西轴线布置吧台及礼宾区，中置岛台书塌，分离出内部交通，西侧吧台背景嵌入由艺术家独立创作的“南京故事”题材浮雕，辟邪、街巷、祥云、秦淮胜境等元素跃然画面，拉开通向民国往事之序幕；浮雕上方正中悬挂，被誉为“当代书峰”乐泉先生创作的草书匾额“红公馆”，取材民国时期保存至今的紫桐木整板雕刻；吧台面放置磁石电话与上方灯盏呼应、书塌上布置百科旧籍、鸟笼、陶罐、烛台等细节，重温历史生活中最细腻部分，也许，公馆的主人正是如此书香世家。客厅东面礼宾区以壁炉为中心，墙面悬挂总统府旧照油画，与西墙面浮雕遥相辉映，粉彩绣墩与提花地毯一副娇滴滴的模样，优雅中透着仪式感。

客厅最为重要的作用除了迎宾送客亦是集散枢纽，南面并置两个入口，分别通往一层堂食区和北二楼的包间区；北二楼布置九个包间，取名“大千食园”“逸仙别院”“美玲客厅”等，分别以民国历史名人为线索取名，包间内布置除基本就餐功能所用的家具外，均以各人物性格展开叙事，还原记忆的画面。通往二楼的楼梯始终暗淡，甚至晦涩，不禁回忆起旧作【竹里馆】中对“通过空间”保有的情感“试图在有温度的交互中保持部分冷静，从而在步入另一个场域前，整理出独立的情绪”。而今，“独立的情绪”只有挂在灰色墙面正中的那幅画与空间和解，画面中推开的窗扇伸向街巷，窗台下粉色的荷花感染着盛夏的余晖，信札刚写了一半便要邮寄出去？似是一个女人的波涛暗涌。正是，灰暗的梯段正是为了波涛暗涌！

The living room works as a transportation hub, which is playing a key role to welcome visitors, distributes them to various rooms and levels. Two side-by-side entrances are open respectively to the dining area on the ground floor and private area accessible to more privacy on the higher level in the north side; there are nine private rooms on the high level, named Universal Pavilion, Yixian pavilion, Meiling Pavilion... which follows a range of well-known names found in the history of Republican China. In addition to the necessary furniture and fixtures catering to the demand of eatery, each of the private rooms intends to tell a story with these features' characters, filming arrays of memorial frames. The stairway leading to the second floor constantly appears dimly dark and obscure. The architect revived a strong feeling from his previous homogeneous project that attempts to keep calm partially while interacting in a warm space and adjust independent mental mood when readily getting access to another distinct room. But now, "free-style mood" could be expressed clearly by the drawing hung on the gray wall peacefully. Windows in the frame extend to the street, and pink lotus flowers under the windowsill are tinted with afterglow in midsummer. Is unfinished letter going to be sent out then? It seems like stormy waves hidden deeply in a woman. Yes, the dim staircases are for the sensitive emotion.

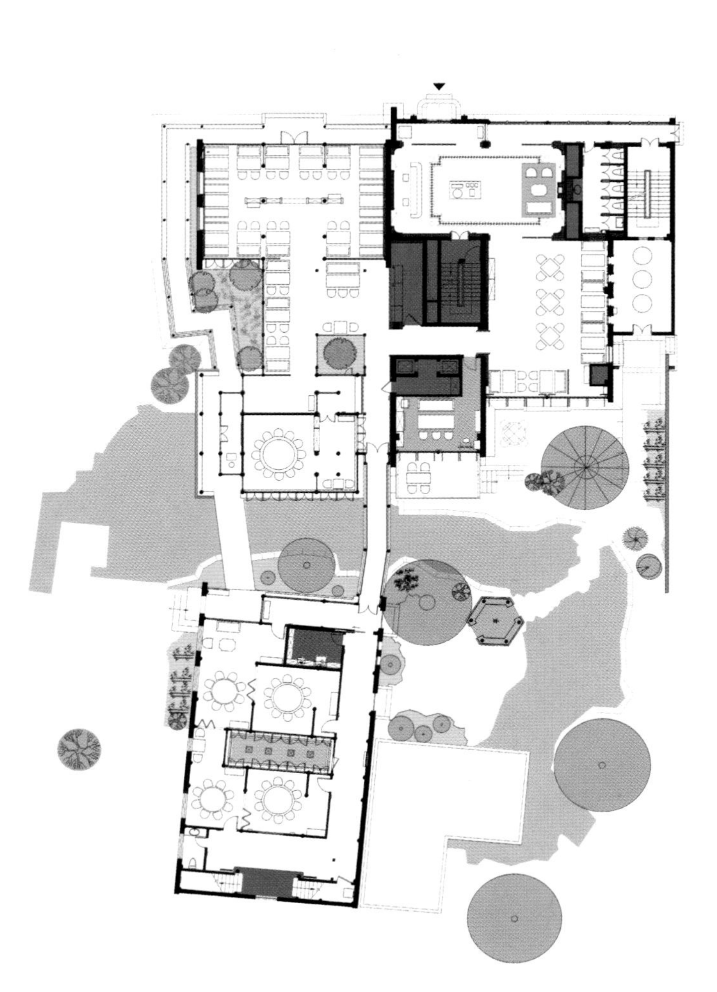

The dining area on the ground floor was separated into two halls, connected by a long passageway. The passageway keeps the principle of 'through space' to get light in the dark. Positive phototropism ensures primarily important dining experience when implementing the rule. The West Hall was transformed from the existing architecture, preserving the main wood structure in the courtyard. An optical center for the area was developed by re-organizing microcosmic courtyard. Meanwhile, reverse supporting structure makes the extension strip off the floor of the existing courtyard to form a much more lithesome volume - as if a modern art installation was set in the classic courtyard. The double-glazed ceiling is used to filter light and save energy at the same time, and luscious, and enough natural light is introduced into the interior when it's sunny day, but artificially projected a beam from outside to light up rooms with beautifully overlapped shadow. It's easy to observe the wonderful scene that dropping water ripples on the surface of the ceiling while it's rainy day.

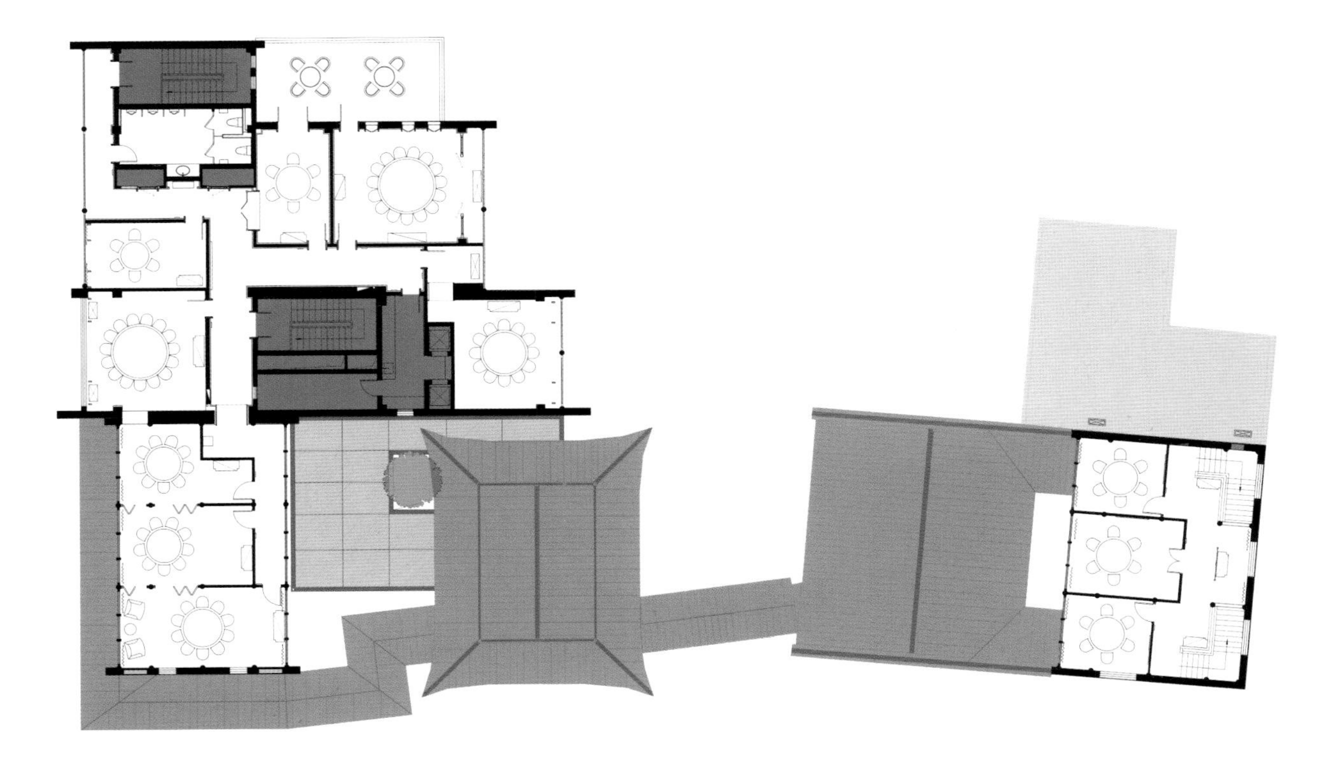

But the pavilion in the south features layout of 'mirrored lobby' originating from the formation of traditional architecture. The corridor in the east orientation was maintained to run through the vertical sides after transformation. However, the rest of rooms comply with structural beams and columns, segmented into four free-standing dining spaces while surrounding skylight inside. The layout of the skylight was re-arranged and back to the basic position of the columns, reformed into a pool of tranquil water waving outside of the building. Some hurricane lanterns float on the water, catering to the interiors harmoniously. Passerby going through and guests taking seats both take an exciting experience of the architecture reflecting on the flowing water, ethereality and tranquility. Oppression and depression from traditional architecture interiors were eliminated ultimately.

Creation

For Jaco, every development means a familiar process of reviewing what has been gained in his past intervention. From the initial brainstorm to feasible modification, an accomplished method belonging to time should be applied in the realm of its technological growth and present it in aesthetic and artistic ways. In his opinion, space is generated according to constraining of a sequence of orders, but detail hides in the logic subdivision. A misunderstood detailing practice by craftsman would march forward beyond detailing creation through racking their brains. It's possible anyway, and it happens again and again in reality, so innovation needs "breakthrough". An idea of systematical construction is much similar to the course of birth and growth of natural creations. The basic gene in such idea should be interpreted and divided properly. Conceived objects succeed conceptual blueprint, which has been confirmed for a long time and evolving to itself wonders.

堂食区分为东西二厅，由过廊相连，过廊保持“通过空间”一贯的营造态度——在黑暗中获取光明，“向光性”是通过空间具备隐晦体验感的保证。堂食西厅由原始建筑院落改建而成，保留院落中的主要树木，通过重组微观庭院形成区域视觉中心。加建部分用反支撑结构将楼板剥离原庭院地面，使之形成更加轻盈的建筑体量，宛若将现代装置置放于古典庭院中。顶面采用双层透明采光顶结构，便于过滤光线与节约能耗，日光下顶面可以获得饱满且充盈的自然光线，夜晚由外部投射照明，光影层层重叠，雨天时可观察到顶面充斥着落水涟漪的视觉奇观。

南楼呈传统建筑形式中“对照厅”布局，改造后仅留东过道为室内交通贯穿南北，其余空间皆遵从建筑梁柱关系，分割为四个独立就餐空间，围天井于内，并以天井平面尺度退让至柱基位置，改造成一池静水飘然屋外，置风灯于水面，与室内交相辉映，行人通过、客人入座，皆可体验到建筑落于水中的轻盈通透，消减了传统建筑室内相对压迫沉闷的感受。

【造】

每一次营造都是温故知新。由意淫到修正，一个时代的营造手法应该尊重这个时代的技术纲领并用艺术的方式呈现出来。空间产生于秩序限定，细节产生于逻辑细分。一个被工匠误解的细部做法有可能超越冥思苦想的细部创造，这是有可能的，并多次出现在现实之中，所以，营造亦需“破执”。一个构筑预想就像自然界中一个完整的生长，需要解读并抽离出最为基本的基因，构筑物存在于构筑之前，一切必浑然天成。

咋

时光印记

THE IMPRINT OF TIME

Project Name | 项目名称
一阑传统牛肉面

Design Company | 设计公司
叙品设计装饰工程有限公司

Designer | 设计师
蒋国兴

Area | 项目面积
500 ㎡

Main Materials | 主要材料
仿木地板砖、白色钢板、白桦树等

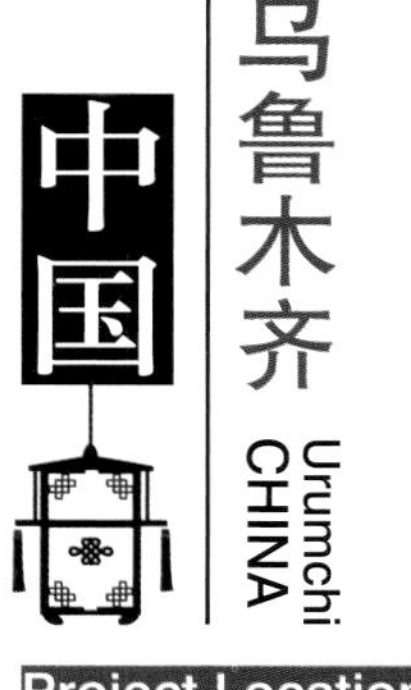

乌鲁木齐
Urumchi
CHINA
Project Location

This case is located in a busy business district. The designer breaking the previous ramen impression creates an extraordinary ramen small pavilion, which has its style with modest and low-key, and makes the taste and vision enjoyment. Time seems to have been driven back into a trance.

The design of door head is particularly important. The iron rust plate used in a large area, white lighting characters and LOGO are a simple and gorgeous modelling highlighting the design feeling in detail. The intersection of light and shades like the invisible hand stroking the annual ring, mottled mark left.

本案位于人车繁忙的繁华商圈，设计师打破以往的拉面印象，打造出一家不平凡的拉面小馆，在朴实低调中有自己的格调，让味觉、视觉得到享受，恍惚时间已经倒流。

门头的设计尤为重要，大面积采用铁锈板，凹凸的木块，白色发光字体及LOGO，造型简单大气，以细节凸显设计感。光影交错，似时光无形手影抚过年轮，留下斑驳印记。

和
揉
溜
剂
拉
品
一蘭

Up to the second floor, the bar uses gray, bright brick which can play a role in stretching the space through reflection. The frames of the bar are decorated with the white steel in ramen shaped modelling, highlighting the subject of this case. Move down inside, all the dining environment is open. In the first individual seat area, the designer uses white birch as the screen to separate the dining area and vestibule area. Facing the original building in a long and narrow type, the designer uses the original roof of industrial wind and the irregular line of white rebar as a ceiling decoration to show a space naturally with a concave surface and in and out routes, and create the continuous of the doorway and the flow of interior space. At the same time, the designer uses the folded field connecting the whole atmosphere of the space.

上到二楼，吧台采用灰色亮面砖，通过反射，起到了拉升空间作用，吧台旁边用白色钢筋作为装饰，造型为拉面状，突出了本案主题。往里走，全开放式的就餐环境，散座一区，白桦树作为屏风，将就餐区和前厅区域分隔开。面对狭长型的原有建筑，工业风的原顶，用白色钢筋不规则的折线作为吊顶装饰，自然表露出内凹的曲面与出入动线的空间，创造出入口的延续与室内的流动，并利用折出的场域，贯穿整体空间的氛围，使机能随形，人处于其中随形而至。

The stable classic tone and solid feeling of the second individual seat area, the metope using rammed earth seems to be plain and simple, and the texture and rough feeling of itself is the best embodiment of nature. Natural hues and classic patchwork are random yet not uncommon, just as we care about the world, but be independent of the outside secular spirit.

The third and fourth individual seat areas, metopes are in a whole modelling, using the gray brick which is the same material with the first individual seat area. The top surface modelling is extended from the first area, connecting the whole space. On the right side of the window, the designer uses white steel hollow lettering with the beam in it, making space a more sense of levels.

散座二区稳定的古典色调和扎实感，墙面采用夯土，看起来朴素平实，本身的肌理和粗糙感便是自然本色最好的体现。天然的色调，经典的拼接随意而又不俗，就像我们关心世界，而又独立于外界世俗的精神。

散座三、四区，墙面造型整体采用灰色条砖，与散座一区选用材料一致，顶面造型由一区延伸过来，贯穿整体，右边的窗户选用白色钢板镂空刻字，投进光束，让空间范围更有层次感。

THE BEAUTIFUL MOMENT OF THE PAST

繁华旧梦

Project Name | 项目名称
大拿炙子烤肉店

Designer | 设计师
孟繁峰

Detailed Designer | 深化设计
席冬

Area | 项目面积
400 m²

Photographer | 摄影师
丛林

In this case, the original building was a 9-meter height slope plant, and then was transformed into a two-storey building by the user which the lowest height of each layer is about 3.3 meters. After the renovation of the designer, it becomes a rich Oriental flavor restaurant.

On the first floor, taking the original load bearing as a center, designers divide the whole layout into a U shape layout. The wine bar is in the location of original load bearing, and the dining area is around it. The kitchen is on the east side of the building, where designers create a door for convenience. The second floor is a back-shaped layout, which is an independent dining area as the focus. This setting ensures the relative independence of each dinning area yet not breaks the shareability of the whole space on the second floor. From the entrance, the aisle is transformed into a circular passage round the inner building. At the end of the south side, there is a pass-through window between the first and second floor. At the end of the north side, there is the only private room of the restaurant, having an independent passage which can directly lead to the parking lot. The private room and the upper part of restroom separate an closed and unavailable space.

本案原建筑是一个9m挑高的坡顶厂房，后由使用者改为两层，每层最低处大约是3.3m。经过设计师的修缮改造，成为一个富有东方情调的餐厅。

一层以原有承重柱为中心将整个布局分列成U字型布局。原承重柱部分为酒水吧的位置，餐位环绕其前后。后厨处于建筑的东侧，开东门便于后厨的进出。二层是回字型布局，其核心是一个独立餐区，这种设置保证每个餐区的相对独立却不打破二层整体空间的共享性。通道从入口便分流，围绕内建筑做环形通道，南侧尽头是一二楼之间传菜口，北侧尽头是全餐厅唯一的包间，包间有独立通道可以直接通往停车场，并和洗手间的上部隔出一个封闭空间，为不开放的区域。

With the red plum blossom as an intent carrier, designers create a spatial conception of quietness, beauty and elegance. On the open vestibule, there is a red plum blossom in full blooming, giving a full of vitality to the whole gray space yet not arrogance and impetuosity. On both sides of the wine bar, there is red plum blossom as segmentation which not only can isolate the sight of diners at dining but also can create a romantic dining environment. Up to the second floor, the segmentation in the building becomes chicer because of the blossoming plum blossom. A beam of light from the east wall shines down over the tip of sticks lightening the prosperity of a tree. Red plum blossom is a soul of the material carrier in this case, so the branches of red plum blossom dissimilate everywhere. The light on the top of the first floor is the transformation of the branches of red plum blossom, and the wall lamp on the brick wall also is. Outside the window, the rhythm of light and shadow comes from the interpretation of the branches of red plum blossom on the steel plate, while it is this branch of red plum blossom that breaks the silence.

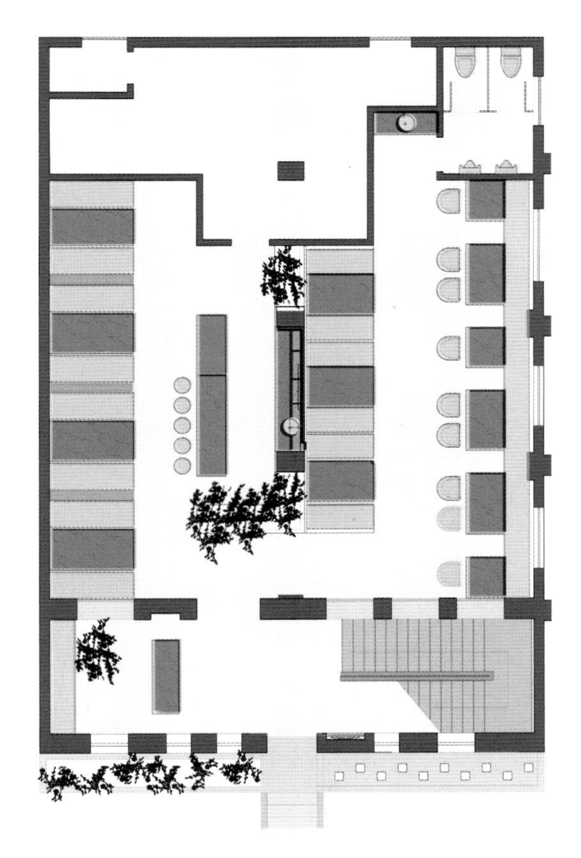

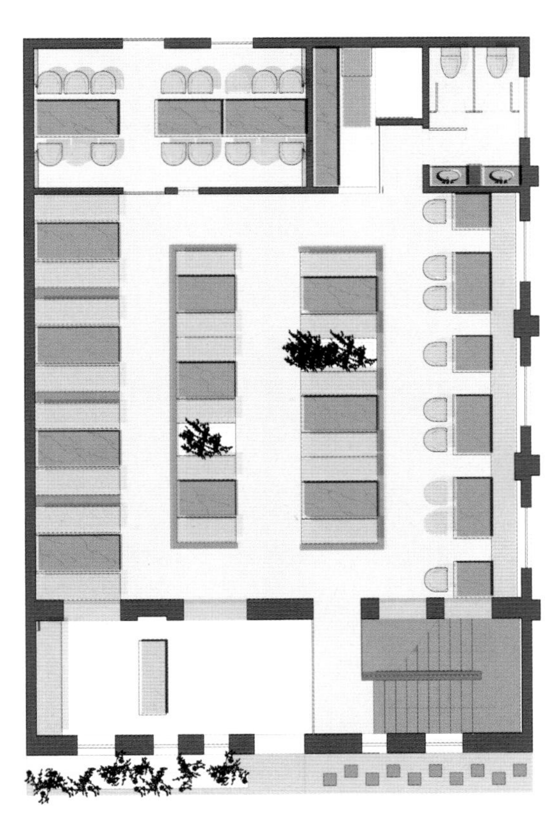

In this case, except beautiful plum blossom, the avant-garde Beauty Painting and extreme abstract concept furnishing accessories also stand out colors. The Beauty Painting of Mr. Hu Yongkai combines perfectly the beauty of lines of traditional Oriental paintings with the color beauty of modern paintings creating a modern Oriental painting and adding more Oriental flavor to space.

设计师借助红艳的梅花为意向载体，营造了一个静美、端庄的空间意境。开敞的前厅有一株怒放着的红梅，让整个灰色的空间陡然充满活力，却不骄不躁，酒水吧的两侧各有红梅做区隔，既隔离了宾客就餐时的视线，也营造出浪漫的就餐环境。走向二层，建筑内的区隔因红梅的绽放而更加别致，东墙一束光打在枯枝梢头，点亮一树繁华。红梅是全案物质载体的灵魂，因此红梅树枝也异化在了各个位置。一层顶部灯光是梅枝变形，青砖上的壁灯亦是如此，窗外光影律动的节奏来自梅枝在钢板上的演绎，墙面打破宁静的仍旧是这枝红梅。

本案承载色彩的除了美艳的梅之外，还有前卫的仕女画和极具抽象概念的陈设饰品。其中胡永凯先生的仕女图画作恰如其分地将传统东方绘画的线条美与现代绘画的色彩美巧妙结合，创造出既富东方情调又具时代气息的画作，为空间增添不少东方意境。

炙子烤肉

CHAOS AND CONFINEMENT

混沌与约束

Project Name | 项目名称
Kinoya

General Contractor | 建筑公司
Pure Renovation

Designer | 设计师
Jean de Lessard

Woodworking | 木匠
Dominic Samson, Solution Durable

Photographer | 摄影师
Adrien Williams

加拿大
蒙特利尔
Montreal
CANADA
Project Location

For its latest Kinoya, the interior designer has tapped into the sources to emulate in his design the primary spirit, function and aesthetics of the izakaya, as the latter was originally an informal place where people drank beer and sake. The transformation is particularly unusual that it explores through extreme design intimacy in relationships between people, making of Kinoya a true representation of the unique approach.

“For a space to become Event or Emotion, it must generate its own energy. I designed an enclosed space that is totally focused on the business of partying. The design elements are deliberately oppressive or aggressive, so that it is anarchic, rough and where we are loudly heckled”, explains Jean de Lessard.

The space, such as how one could figure what the interior of origami looks like, is composed of triangles of various sizes, crookedly placed in a random fashion. "Jean told me what he wanted to feel in this place. Where one had to be cramped also. It's a fantasy cave where people are in a constant visual exploration mode", says artist carpenter Dominic Samson, Solution durable, who built the structure, a piece of work he’s proud of and that he describes as uplifting.

祭
うどん
うどん

居酒屋
焼きとり
うどん
焼きとり
祭
Bar
Restaurant

在Kinoya的设计上，室内设计师从源头上模仿居酒屋的主要精神、功能和美学理念，因为居酒屋原本是一个人们喝酒消遣的场所。这种转变是非同寻常的，它是通过极端设计来探索人们的亲密关系，用独特的手法对Kinoya真实再现。

“一个空间无论最后成为一个项目还是一种情感，它都必须产生其自己的力量。我设计的这个封闭空间是针对于商业聚会的，因此我故意采用了一些极具压迫感和侵略性的设计元素。所以它是混乱的、粗糙的，人们可以在其中大声质问对方。”Jean de Lessard解释道。

从如何描绘出折纸内部看来，这个空间是由不同大小的三角形以任意形式弯曲地放置在一起。“Jean告诉过我他想在这个空间感受到什么。在这里，人们也需要受到约束。这是一个幻想的洞穴，人们一直处于这种视觉探索模式当中。”木匠Dominic Samson说道。他提出的解决方案十分耐用，这是一件让他骄傲，令人振奋的作品。

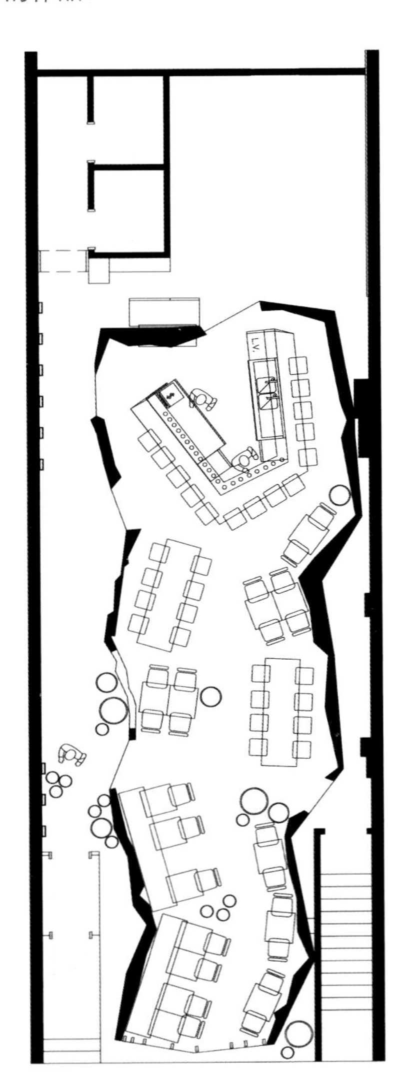

祭
うどん
祭
うどん

焼き
とり
うどん
祭

The reused wood from barns is local and covers an area that represents 4,500 square feet. Durable material, wood has an exceptional capacity of resonance and absorption. The irregularity and angularity of the surfaces further deflect sound waves, helping to muffle the ambient noise. Boards of hemlock and white spruce of different width and thickness were installed in all directions. The vertical drop of 4 to 5 feet between the front and rear parts of the ceiling contributes to the cocoon effect. If this strengthens the idea of chaos, on the other end the glued-laminated technique used for the installation provides, in turn, a perfect finish.

The place is always full since the opening, despite the fact that one must stand shoulder to shoulder. The soft lighting and the cozy atmosphere makes it a friendly environment where the smell of wood mingles pleasantly with the aromas of mouth-watering dishes.

In Japan, an izakaya is a place of socialization and of stress alleviation. Here at Kinoya, the narrow space forces to relate to one another, under his/her unavoidable gaze. The design has the West and the Far East (East Asia) beliefs about community spirit, closeness and brotherhood collide in fun and joyful manner.

这个可重复使用的木材来源于当地的谷仓，占地面积为418m²。它是一种耐用材料，具有超常的共振吸收能力。其不规则的表面，有利于声波的转移，因此其有较好的隔音效果。不同宽度和厚度的铁杉和白云杉板均安装在各个方向。前后天花板的10～12cm的垂直落差有助于形成一种虫茧的效果。这样的设计一方面是为了强调一种“混乱”的理念，另一方面使用迭片黏合技术，使其拥有一个更完美的外观效果。

自从开业以来，居酒屋的客人一直都很多，必须摩肩接踵。柔和的灯光和惬意的气氛给予人们一种亲切感，时不时还有各种美味佳肴和木材本身的味道袭来。

在日本，居酒屋是一个缓解压力的场所。在Kinoya，在狭窄的空间压力下，四目相对不可避免。这个设计传递出将西方和远东地区关于团结和亲密的信仰以一种娱乐和有趣的方式相碰撞。

A RELAXING STATION AT NIGHT

夜幕下的栖居站

Project Name | 项目名称
Rico de Kitchen

Design Company | 设计公司
Yusaku Kaneshiro+Zokei-syudan Co.,Ltd

Designers | 设计师
Yusaku Kaneshiro, Masako Suzuki

Area | 项目面积
83 m²

Photographer | 摄影师
Masahiro Ishibashi

It's a stylish and warm beer cafe whose theme is beer barrels and flowers. The restaurant uses the beer barrels as the main element, and piles them up into distinct arch shapes. Through the connection of flowers, vines and beer barrels, under the mapping of lights, the distinctive style features of Japanese Bistro are created. At the same time, we decorated and divided the customer's seats by the arch piled with beer barrels, and decorated the whole space with flowers, patterns, and words. While the wooden frame of black and white fabric seats echoes with the tone of beer barrels. The geometric patterns on the fabric concisely highlight the modern fashionable texture and render an active atmosphere.

The favorite is the iron garland lamp above the restaurant, which the light surrounded by flowers and vines is like the starlight. This is a romantic space, created to attract women in the way a cafe can be. When darkness falls, turning on the lights of the tavern, the whole restaurant gives a sense of warmth and closeness in the rendering of the light effect. After a day of tired and busy, customers can have a drink with several friends, empty the exhausted mind, and enjoy the life.

RICO
DE
KITCHEN

RICO
DE
KITCHEN

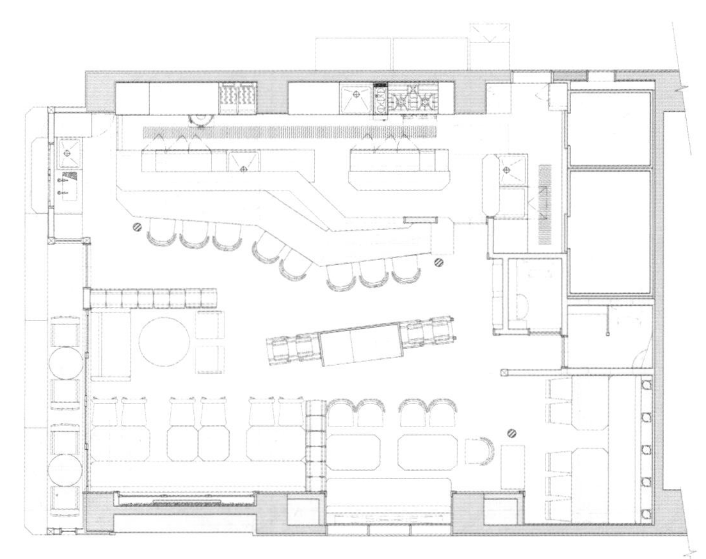

这是一家以啤酒桶和鲜花为主题的啤酒小餐馆，格调别致温馨。店内以酒桶为主题元素，堆砌成造型各异的拱门形状。通过鲜花藤蔓与啤酒桶的相互串连，在灯光的映射下，打造出日式小酒馆鲜明的风格特色。同时，我们用啤酒桶堆出拱形门来装饰与分隔座位区，用鲜花、图案、文字等来装饰整个空间。木质架构的黑白调布艺座椅，与啤酒桶的色调形成和谐呼应。布艺上线条各异的几何图案花纹，简洁明了地突出了现代时尚质感，活跃了空间氛围。

RICO
DE
KITCHEN

RICO
DE
KITCHEN

餐厅上方的铁艺花环是最受人喜爱的，灯光在花朵与藤蔓的簇拥下，星光点点。这是一个浪漫的空间，设计师创造这个空间是为了尽可能地吸引女性顾客。夜幕降临时，把小酒馆的灯光打开，整个居酒屋在灯光效果的渲染下，给人亲近温馨之感。在一天的奔走劳累过后，拉上两三好友，放下疲惫的心灵，一定要进去喝一杯，方才快意。

CAFE & BEER
RICO DE KITCHEN
CAFE & BEER
RICO DE KITCHEN

RICO DE KITCHEN
CAFE & BEER
RICO DE KITCHEN

THE MEMORY TASTE OF THE PAST

久远的记忆味道

Project Name | 项目名称
渝江记忆火锅

Design Company | 设计公司
郑州青草地装饰设计有限公司

Area | 项目面积
500 ㎡

Main Materials | 主要材料
老榆木、麻绳、红色玻璃等

Photographer | 摄影师
吴辉

The things in the memory which are historical, are always dim like the hotpot in the memory. Chongqing Memory Hotpot is a historical hotpot. The memory is always very far away. The gray ground, old elm of half wall and the random but ordered hemp rope resemble the time and space in that memory with a lasting taste, while the weak red lights from pieces of red glasses render a sense of times.

记忆中的东西，有历史的东西，总有些暗淡，如同记忆中的火锅。渝江记忆老火锅，是有历史的火锅。记忆总是很遥远的东西，灰色的地，老榆木的半墙，随意却整齐排列的麻绳，就如那记忆中的时空，有着久远的味道，却也因片片的红色玻璃，弱弱的红光，多了一些时代感。

The designer uses the hemp ropes as the wall to maximize the limited space. Here not only has the pure of hotpot but also has the delicate space. The partition and ceiling of hemp ropes through the lights leave the mottled shades to render a sense of funny. Several friends get together in this space where appears indistinctly with a popular song. Poking a hemp rope, life is showed everywhere.

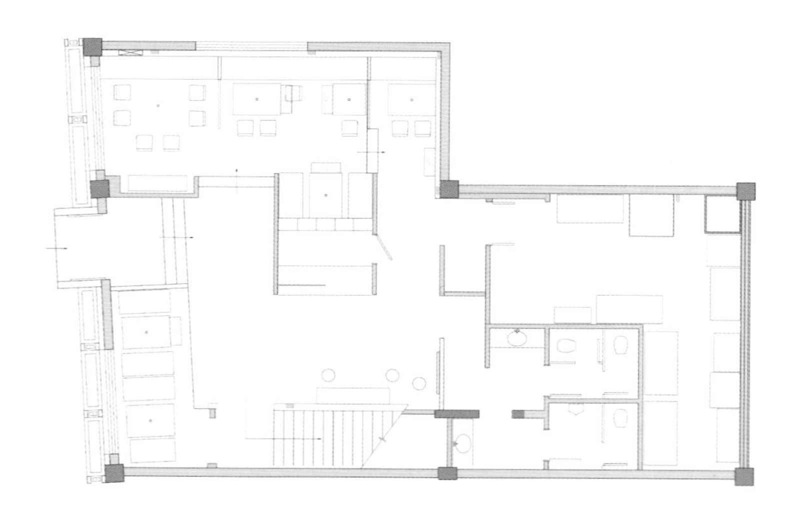

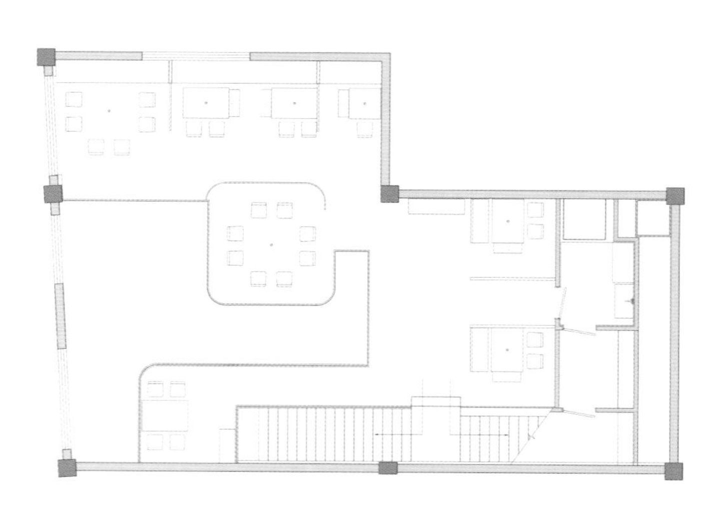

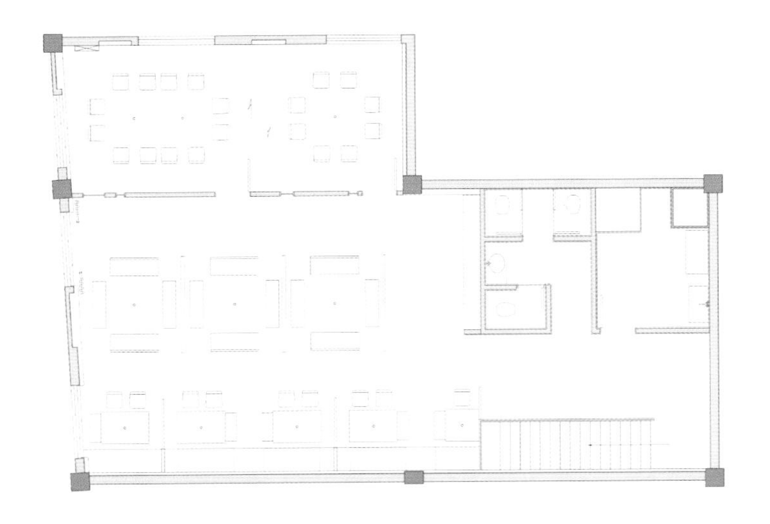

以麻绳为墙，将有限的空间无限化，错落着点点的红灯，满是画面，看不穿的尽头。这里不仅是火锅的纯粹，更有空间的极致。麻绳的隔断，麻绳的天花，透过的灯光，斑驳着影子，多了一些乐趣。若隐若现的空间，三五好友相聚于此，满满的美食，配上一曲火火的音乐，谈笑间拨开一撮麻绳，随处都是生活。

AN ETHEREAL SILENT SCENE

空灵寂静景，深存敬畏心

Project Name | 项目名称
轻井泽拾七石头火锅永春东七店

Design Company | 设计公司
周易设计工作室

Chief Designer | 主持设计
周易

Participant Designer | 参与设计
杨淙琦

Area | 项目面积
1480 ㎡

Main Materials | 主要材料
文化石、铁刀木皮染黑、锯纹面白橡木皮、杉木实木断面、旧木料、黑卵石等

Photographer | 摄影师
和风摄影吕国企

Dining is relaxed or ceremonious. In order to pay tribute to the ingredients, nature and excellent cooking skills, maybe we can regard the dining space as a kind of enchantment full of spirit. This case is hoped to increase the dining process and five senses to a higher level through the presentation of design, context and vocabulary.

The building moves inward nearly eight meters from the roadside for giving four meters to waterscape and four meters to waiting area. This abundant space combined with low tables, plant low hedge, lighting design, Zenist waterscape and the control on horizontal visual line, successively shows the transformations of views of each layer from outside toward inside, and delivers a containing concept on boundaries.

用餐，是一种可以很放松、也可以很隆重的仪式。为了向食材、天地、精湛的庖工技艺致敬，不妨将用餐空间视作一种充满灵性的结界。本案设计师正是希望透过设计、情境、语汇的铺陈，将用餐的过程、五感提升到更高层次。

建筑自路边起向内退缩近8m，让出4m水景加4m等候区的充裕纵深，结合低台度、栽种矮篱、灯光设计、禅意水景以及水平视线的掌握，由外向内层层景色起承转合，传递内敛的界线概念。

拾七

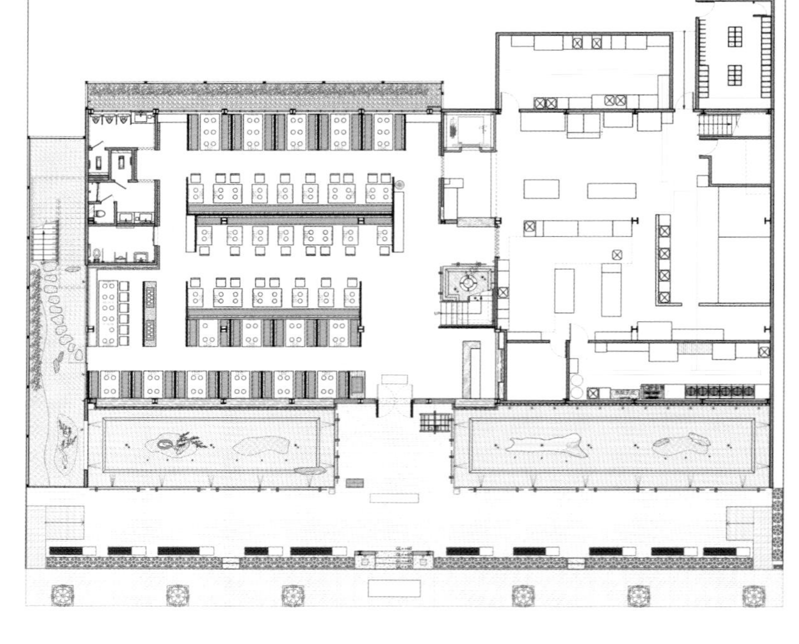

A huge black iron arch stands on the central entrance to divide indoor and outdoor area, and the power inscription of Chinese characters echoes with the towering building. The facade of the building is equipped with a silent double-pitched roof to highlight the traditional Japanese style's profile line which is a close fusion of white mud, rammed earth and wooden pole. Under the eaves, the huge Shimenawa uses FRP material to simulate the symbolization of God's enchantment. Under the contrast of lights, the entangled hemp fiber texture is extremely realistic, which not only can weaken the presence of a busy world, but also release a mysterious atmosphere with emptiness and silence.

Inspired by Japanese shrine, the exterior of the building is a reworking of the quiet aesthetic in ancient buildings, using the matching board and black tiles to outline a simple and rusty iron-like texture feeling, showing a flavor of quietness and solemness. Around the waiting area expanded horizontally, cool water landscapes are arranged along both of the two sides. The pool is inlaid with burned trees and towering landscape stone which metaphors a floating island. The wood burned emptied forms a delicate waterway with the gurgling water sound, and greensward contrasts the landscape stone creating a delicate picture for viewing.

Entering the door, there is a counter built with thick-solid rock and wood. The background bamboo weaving echoes with the shades of light like the crooning whisper. On the first and second floors, the customer seats are arranged with deep and quiet views, surrounded by warm and quiet wood tone without any hubbub. Lightweight wood grid grows and expands from top and bottom, and respectively transplants the nostalgic feeling of Showa period in Japan to achieve the goal of multi-layer demarcation and the beginning of the background light. The drum lamps are hanged over the customer seats, and the faint light renders a touch of romance. One dynamic and one static contrast the shadows of people coming in and out of the space.

On the back of the first floor, space is reserved for building a one-meter-deep patio for lights, and green bamboos are planted to decorate the place. When the lights are on in the night, the place becomes a dream-like scenery. Climbing up the stairs, through the lights projecting, the overhanging deadwoods depict ever-changing splash-ink in the wall, having the fun of one view in one step, forming a special format of showing welcome. The local staircase wall broad the vision by iron mesh, branded a mobile glimpse. On the second floor, the customer seats expand and continue the parallel axis, and the wood grill surrounded on those seats emerge the solemn imagery. Both sides of the aisle in the last section are placed with Buddha water feature and stone counters. Mood is created by heart, but people all are so.

中央入口以巨大黑铁牌楼象征里外，笔力遒劲的汉字与巍峨的建筑相呼应。建筑正面铸以静穆双斜顶，彰显传统日式民居惯以白泥、夯土、木柱紧密交融的剖面线条，檐下以FRP材质模拟象征神灵结界的巨大祝连绳，在灯光烘托下，交缠的麻纤肌理极其逼真，不仅能弱化车水马龙的存在感，更释放出一种空灵和寂静俱在的神秘氛围。

建筑外观是古建筑静谧美学的再淬炼，用企口版与黑瓦堆栈，勾勒古朴粗糙的类锈铁质感，表现静穆之韵。横向展开的等候区，两侧设计沁凉水景，池间镶嵌火炽枯木和隐喻浮岛的峥嵘景石。被烧空的木头形成精巧水道，水声淙淙，翠绿草皮烘托景石，营造出袖珍画境。

进门是厚实岩块搭配实木台面砌成的柜台。背景的竹编肌理呼应灯光的明暗，仿佛低吟絮语。一、二楼设置景深幽邃的客座群落，环顾尽是温暖安静的木头基调，没有丝毫喧哗，半人高的轻盈木格栅分别自天、地发芽、伸展，完成多层次的划界目的，也成就背景光的滥觞。客座上方悬挂着鼓灯，幽微的光，兀自晕着久远浪漫，一动一静，对比着不断来去穿梭的人影。

一楼后段预留1m纵深打造采光天井，翠绿修竹掩映微风间，娉婷摇曳的丰姿陪伴用餐。夜里华灯初上，更有如梦境般的旖旎。拾阶而上，悬垂的枯枝经过灯光投影，在素墙上留下瞬息万变的印象泼墨，有一步一景的趣味，成为迎宾的另一形式。局部梯间墙面以铁网格窗打开视野，烙印移动间的惊鸿一瞥。二楼客座延续并列轴线，上围框绕周边的木格栅，让空间涌现庄严的意象，末段的过道两侧分置佛陀水景与石砌柜台，境由心生，但凡如此。

THE FLAVOR OF TEA

茶香水云间

Project Name | 项目名称
水云间 · 茶会所

Design Company | 设计公司
苏州叙品设计装饰工程有限公司

Chief Designer | 主案设计
蒋国兴

Area | 项目面积
460 ㎡

Photographer | 摄影师
吴辉

Main Materials | 主要材料
黑色花岗岩、火山岩、海藻泥、鹅卵石、米黄洞石、白色乳胶漆、水曲柳做旧木饰面、竹帘、实木复合地板等

Life is like a cup of tea, which the first one is bitter as life, the second fragrance as love and the third bright as a breeze. In this rapidly developing and noisy city, the tea room is the first choice for seeking a real peace of mind. In this case, the modern Chinese style is used to present a simple, elegant and chic tea-culture space. There are two storeys in the plane layout, the first floor including the hall, reception area, private rooms and landscape area, and the second floor including four small private rooms, one large private room, kitchen, restroom, office and storage room, etc. In the use of colors, black, white, do-old wooden colors are dominated as the main tone while the gray color as the auxiliary tone.

Entering the vestibule, there is a neat cement pile wall fulfilling with candlelight in every hole, exuding a partly invisible and partly visible light. Through the rust plate sliding door, there is the hall. The irregular Chinese characters "水云间" sculptured by rust plate hanging on the wall made of volcanic. The cabinets in walls are divided into frames in all sizes, which every frame is divided into some small boxes like the small cabinets of ancient pharmacies. A variety of small pots are placed in the concave grids and seems more delicate in the yellow light.

At the back of the vestibule, there is a reception area without any luxurious decoration, while it blocks the interaction with hall by a mud wall. There are landscape area and private rooms on the right side. Whether you are in the outdoor or the entrance, reception area, hall and private rooms of indoor, you can enjoy the landscape decorated with sticks, stones and cloud lamps as an accent. There is no complete separation between private rooms and hall, and the glass partition contains the visual contact with the hall. The wall decorated with volcano rock hanging bamboo fence, the cabinet door with wooden latticework, the black and white paintings, all of these elements reveal a Chinese style.

一茶一世界，一味一人生。人生如茶，第一道茶苦若生命、第二道香似爱情、第三道茶淡如清风。在这个飞速发展、喧闹的城市中，寻求内心一份真实的平静，茶会所便是首先。本案用现代中式风格展示了一个素雅、别致的茶文化空间。在平面布置上分为2层，一楼规划了大厅、接待区、包间、景观区，二楼规划了4个小包间和1个大包间、厨房、卫生间、办公室、库房等。在色彩运用上，以黑色、白色、做旧木本色为主色调，灰色为辅色调。

进入前厅是水泥块整齐堆砌的背景墙，每一个洞口都摆满了蜡烛灯，散发出若隐若现的灯光。穿过铁锈板的移门，便是大厅。锈铁板雕刻的不规则“水云间”字体悬挂于火山岩墙面上。墙面的柜子被分成了大大小小的框架，每一个框架又被等分了许多小格，很像古代药房的小柜子。内凹的小格子摆满了各种各样的小陶罐，在淡黄色的灯光下愈发显得精致。

前厅的后面是接待区，没有奢华的装饰，而是以一面残岩断壁土墙阻挡了与大厅间的交流。右边是景观区和包间，枯竹、石头、云灯装饰的景观，无论你置身在室外还是室内的入口处、接待区、大厅、还是包间，都能欣赏到，可谓点睛之笔。包间与大厅之间没有完全隔开，玻璃隔断很好地保留了与大厅间的视线交流。火山岩装饰的墙面挂着竹篱笆，木制花格装饰的柜门，黑白的挂画，无一不透露着中式情节。

On the entrance to the stairs, the designer using Chinese classical elements designs an octagonal partition to show smooth and slick in life by adopting circuitous progressive design techniques. Across the octagonal partition, a long table stands there. Under the desktop, an irregular iron rust plate with several poems at its back, adds a bit of colors to the black ground through the glossy black stone reflected on the ground. There is a ceiling lamp decorated with bamboo poles hanging on the top of the table, which is special and practical. Passing the partition wall, a giant pool is placed in the center, one side is the entrance of stairs which straightly leads to the second floor and a floating cloud on the top means the success in business, while the other side is a zither-playing area suspended on the pool. Many candle lights are projected in the pool. The weak light emitting into the water with a zither-playing sound seems harmonious and beautiful. The stairs to the second floor are concise and practical. The black stairs, white railings, bronze handrails, hidden lights, the cloud-shape lamp on the top, and the Xu Zhimo's poetry "Very quietly I leave, as quietly as I came here" on the wall, all of these sublimate the concept to a proper extent.

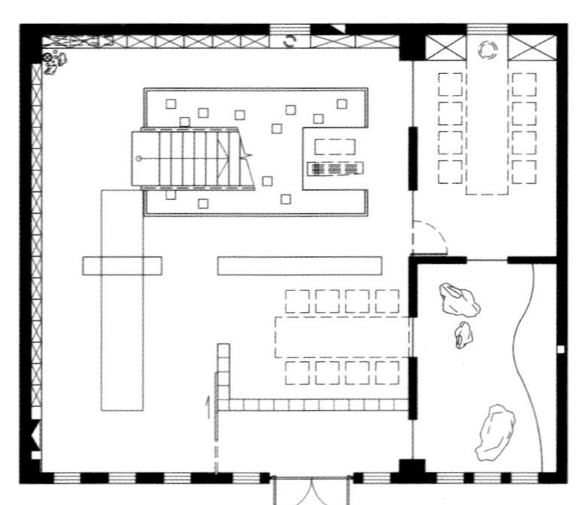

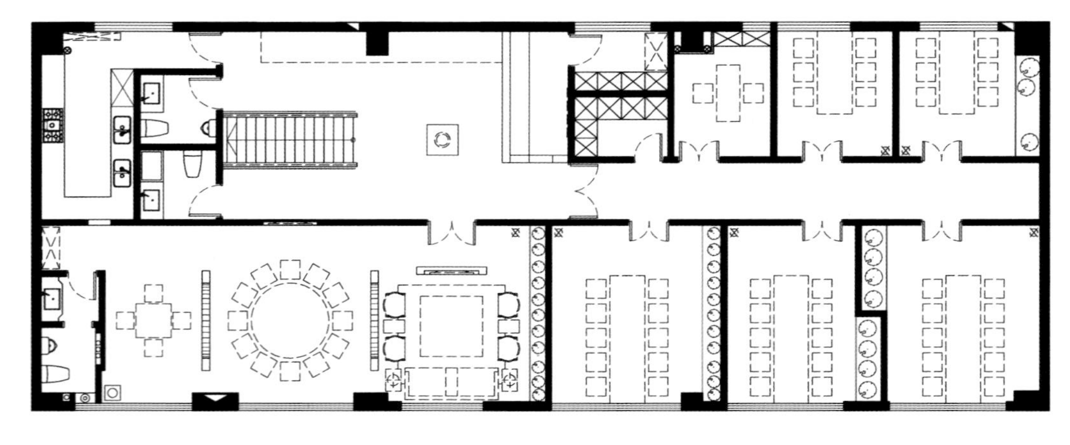

On the left of the second floor, there is the landscape decorated with bamboo, and the reception desk and the whole walkway decorated with slats. There is no extra light in the aisle which is divided into several parts by three back-shaped lamp slices. The entrance of each blank door is hanging on a doorplate sculptured by rust iron plate which is shined by a bunch of light. At the end of the corridor, a cloud-shaped lamp is an embellishment of the whole aisle, and the mirror in the wall stretches a sense of space.

The private room is dominated with plain color without any complicated models. The walls are decorated with beige travertine, pebbles, gray slats and gray volcanic, allocating with the decorative accessories like bamboo fence, black and white paintings and branches. There is a bamboo landscape in each private room which can make the space more romantic. The restroom continues the elements of corridor which the walls decorated with slats and the black ground make space more rustic.

在通往楼梯的入口处，设计师运用中国古典元素，采用迂回渐进的设计手法，特意设计了一个八角隔断墙，意味着人生的八面玲珑。一条长长的桌子穿过八角立在那里，桌面的下面是一块不规则的锈铁板，背面写了几首诗词，通过黑色亮面的石材反射到地面上，给黑色的地面增加了一点点色彩。桌子的上面悬挂着一根竹竿装饰的吊灯，既特别又实用。绕过隔断墙，一个巨大的水池置于中央，一边是楼梯的入口，直直地通向二楼，顶面还飘来一片云朵，意味着生意上的平步青云，另一边是悬空于水池上方的古筝演奏区。水池中规划了许多蜡烛灯，柔弱的灯光散发在水面上，再配上古筝的演奏声，既和谐又美好。通往二楼的楼梯简洁实用，黑色的踏步、白色的栏杆、古铜色的扶手、暗藏的灯带和顶面的云灯，墙面徐志摩的诗词：“轻轻的我走了，正如我轻轻的来”，恰如其分地升华了意境。

二楼左边是竹子装饰的景观，木拼条装饰着服务台及整个走道。走道没有多余的光源，三条回字形的灯片把过道分成几段，在每个暗门的入口处都悬挂了一个锈铁板雕刻的门牌，一束束光源照射在每个门牌上，在走道的尽头，一朵云灯点缀了整个过道，墙面的镜子拉伸了空间感。

包间没有复杂的造型，以素色为主。米黄洞石、鹅卵石、灰色木拼条、灰色火山岩装饰的墙面，再搭配竹篱笆、黑白挂画、枯枝等装饰品。在每个包间都规划了一处竹子的景观，使空间更具有情调。卫生间延续了走道的元素，木拼条装饰的墙面，黑色的地面使空间看起来更质朴。

莲说

ODE TO A LOTUS

Project Name | 项目名称
苏州量子 · 馋源餐厅

Design Company | 设计公司
FCD 浮尘设计

Designers | 设计师
万浮尘、唐海航

Area | 项目面积
600 m²

Main Materials | 主要材料
水泥、复合木地板、槽钢、钢丝、装饰木条等

Photographer | 摄影师
潘宇峰

This case is located on the east side of Xietang Old Street of Suzhou containing 900 square metres area. It is a catering enterprise specializing in the local characteristic diet culture of Suzhou. According to the promotion idea of Modernist Cuisine · Chanyuan Brand - the theme of “Eating · Life”, designers combine with the unique historical cultures and regional characteristics of Suzhou to extend its brand culture through the overall creative design of space.

本案例地处苏州，位于斜塘老街的东侧，建筑面积约900m²，是苏州本地专门研究苏州当地特色饮食文化的餐饮企业。设计师根据量子 · 馋源品牌的推广理念：以“吃 · 生活”为主题，结合苏州特有的历史文化和地域特色，通过空间的总体创意设计，将其品牌文化进行了进一步的延伸。

消防栓

馋源
WELCOME
量子 馋源
TIME
下午 17:00~20:30
0512-67678979
WIFI

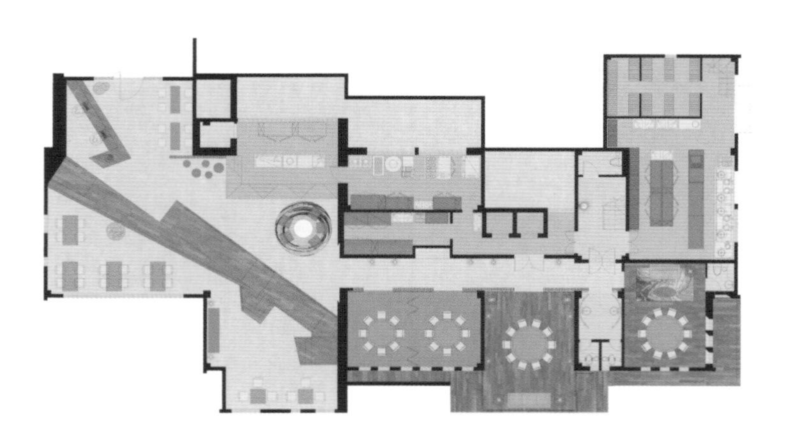

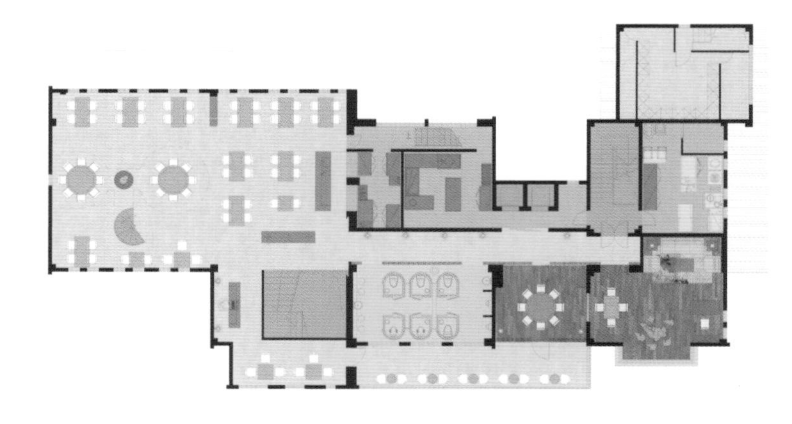

This case is based on the "lotus" as the main design element. Designers integrate the understanding of the shape and meaning of "lotus" into the design concept of space with a full extension and extraction. The original lotus-petal-shape seats, lotus-shape chandeliers and decorations can be seen everywhere in the whole space. The whole space is based on the dark color, implanting the silver and reflective "fishing net" made of steel wires to create the Suzhou characteristics as an abundant place. At the same time, designers take advantage of floor height to create a local split-level structure. The lotus chairs placed on the local split-level structure and the silver steel wire gauze below it through the light exposure, present a beautiful scene which is dining on the fishing net with lotus. While the design of a large area of cement wall is plain and comfortable with a touch of Zen flavor. This case highlights the cultural theme of "Eating · Life", emphasizes the awareness of brand and promotes the charm of Suzhou diet culture.

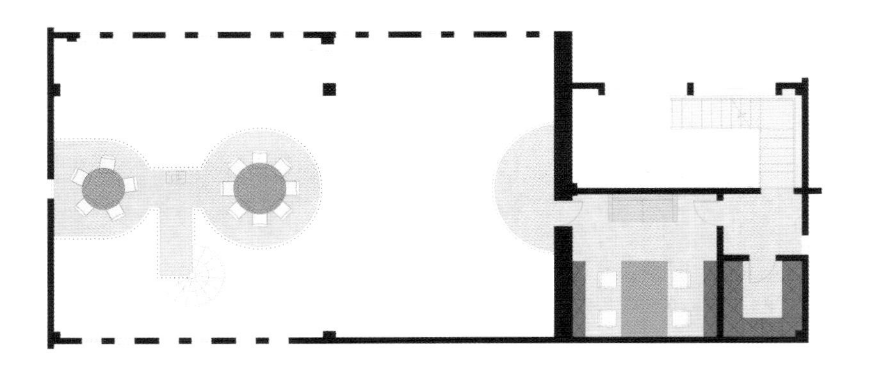

本项目以“莲”作为主要的设计元素，通过对“莲”形与意的理解，融入到空间的设计理念中，做到充分地延伸和提炼。如整个空间区域内随处可见的原创莲花瓣座椅、莲花吊灯及装饰品等。整体空间以深色调为主，在空间中植入钢丝编织成的银白色具有反光的“渔网”，营造出苏州鱼米之乡的特色。同时利用层高的优势，做了局部的错层结构，错层上面摆放的莲花餐椅，下部有银白色的钢丝编织的钢丝网，通过灯光的照射，给客人呈现出一副在渔网之上，坐拥莲花用餐的美好景象。而大面积的水泥墙面设计，质朴舒适，又带着淡淡禅意。以突出“吃·生活”的文化主题，强调品牌意识，弘扬苏州饮食文化的魅力。

浓情诗饮

GREAT AFFECTION IN DRINKING POETICALLY

Project Location

"I offer tea, not wine, to a guest on a cold night; When water boils on the stove, fire burns with flame bright." This is the Song Dynasty poet Du Lei described that master fired furnace and boiled tea, with tea to receive guest hospitality on a cold night. With a bright fragrance of warm tea, presenting their great affection during the conversation and tea-tasting, this kind of elegance was a leisurely elegant charm conveyed by Song Dynasty people, making the later generations long for this kind of elegant life. There is also someone wants to taste the boundless lonely in this contemporary time. This is the reason why the Bamboo's Eatery was created.

Project Name | 项目名称
竹里馆

Design Company | 设计公司
南京名谷设计机构

Designer | 设计师
潘冉

Area | 项目面积
900 ㎡

Main Materials | 主要材料
竹、泥灰、木板等

This is a three-storey building facing the local street exactly with intervention of Dissipation philosophy and vessels made of plain bamboo implying modest persons. The architect tended to fulfil a kind of establishment ideally. The establishment is proximately civil engineering in folk, which is not subject to serious architecture. The emerged experience generating from the built space just placed “dissipation” into the re-arranged spatial order, which “light” is the most important elements in this order. And also it can be recomposed by the “dissipation” from the built space to gain the double emotions of light and space. “Dissipation” can tell you how to shape an elastic light! If “the built space” is an attempt to relax, the carding is a complete rational analysis.

Linear extension spreads alongside the facade to the doorway in the main entrance, broken into lateral distributary and leads the way to the Tea Break Area of Floor One deliberately. This area was embroidered with homogeneous latitude on the formation of semi-space to restrain the section. Tea seats were set alongside the bamboo fence to develop a continuous two-party space relation, thus composing the core function zone of Floor One—“Garden Fence”. Meanwhile, the bamboo fence surrounding “Garden Fence” transforms latitudinally and plays the role of guidance to connect the bar, dishes’ review and service, linking the core zone precisely and closely, and ultimately leading to the going-up elevator.

Walkway to Floor Two is connected by step staircases cladded with steel plates, attempting to keep calm partially when interacting with warm atmosphere and standing for a narrative to set up an independent sensation before going into another area.

Tea Break area on Floor Two approaches windows strategically and appears harmonious visually. The customers are more easy to feel the poetry scene which is the light shines on the table through the window. Side to the South was segmented into tea seats and elevator lobby by applying interweaving bamboo arrays horizontally and perpendicularly. Bamboo arrays extending to the east end of the toilet and lift up the water tank. Water floats down the stream through bamboo-made tubes, which is carefully planned to weaken the heaviness of water tank.

“寒夜客来茶当酒，竹炉汤沸火初红。”这是宋代诗人杜耒描写在寒冷的夜里，主人点炉煮茶，以茶当酒待客的诗句。清香茶暖，品茗交谈中其情浓浓，此中儒雅正是宋人传递出的悠悠风韵，令后世神往的高雅生活。当代浮世尽欢，亦有静心品味当下无边落寞者，竹里馆为此而立。

一栋三层临街小楼，以魏晋消散之气为道，喻意君子的白竹为器，尝试一种搭建。搭建似乎更像游离在严肃建筑学之外的民间土木，而搭建带来的空间体验正是将“散”放置在被重新梳理的空间秩序中，这种秩序里最重要的因素 ——“光”亦是被搭建所带来的“散”重新分解，而获得光线与空间的双重情感，“散”可以告诉你如何塑造弹性的光线。如果说“搭建”是一种放松的尝试，那么梳理则是完整的理性分析。

由外立面的竖向线条延伸至主入口玄关，形成侧向分流进入一层茶歇区，将竹用单一纬度的围合方式形成半空间限定区间，茶座布置在竹篱一侧，形成二方连续式的空间关系，并由此聚合成一层的功能核心 ——“篱园”。此时，围绕着“篱园”的顶面竹篱正发生着纬度关系的转变，并引导性地将吧台、出品、服务动线等功能串连起来，与之前的功能核心形成咬合关系而最终指向通向上层的垂直电梯。

通往二层的交通增加了北边的步行体验式楼梯，氧化钢板制作的梯段，尝试在有温度的交互中保持部分冷静，从而在进入另一个场域前，以一种旁白的姿态重新整理出独立的情绪。

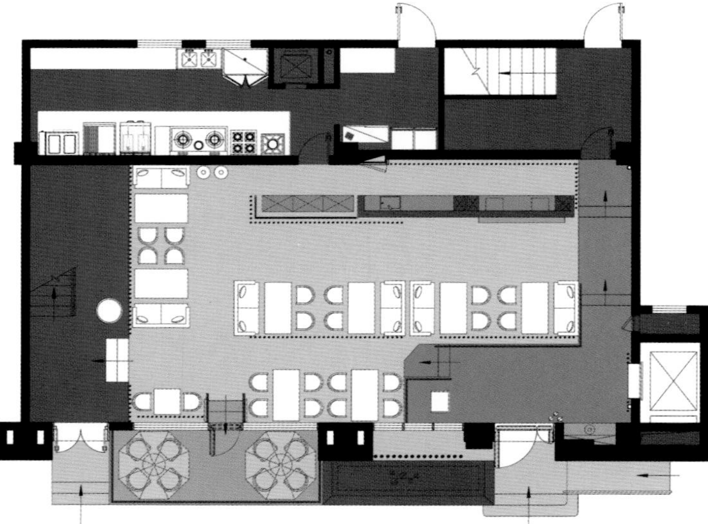

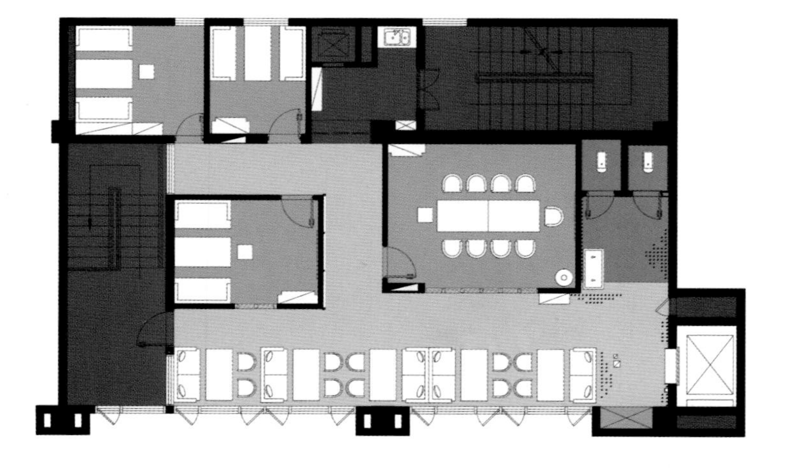

The entrance of the private rooms on the second floor is contained in a relatively compressed volume, and the "oppression" is for a better "release". In the condition without natural lights, the designer takes the wall of the western compartment and public area to introduce the light by boring the holes. The natural light through tea break area transfers to the private rooms without mottled moving lights, but it is warm and bright. The filtered light in the daytime is like a net that produces the light in the relatively dark space. In the corridor of the private rooms, the walls are filled with warm white putty, and there are also placing historical wall tiles with seven hundred years, which have the shuttle experience of time walking in it. The inner wall of private rooms with lots of white spaces takes the attitude of simple and plain to leave an interface of enough imagination for literati.

All rooms on the third floor are free-running teahouse in the east and the west. Walkway centralizes the entire level and makes a turn slightly in the south and the north. There stands a 'thatched cottage' architecture, which its eaves hang down and bamboo windows are raised up from the inside out. Whether you are inside or outside, there seems to be a desire to check it out from the window.

In the long and narrow walkway, with the purpose to obtain walking experience of "exploration in tranquility" and introduce natural light effectively into a whole sealing space which is departed from direct daylighting, artificial installation called "Liangzi" was introduced thereby, which can solve the exchanging light problem in isolation. Therefore, bi-layered daylighting interface overlaid by bamboo fence upside the shadow walls becomes a warrior to lighten darkness temporarily. The local construction of the top turning point treated by sculpture is the wanton gesture after the liberation of light in space, and effectively softens the relatively tough space docking. The private rooms are still simple and plain with a lots of white spaces.

Sitting down, reminding of seven sages of the bamboo grove and Tao Yuanming in the farming and reading, maybe it will be completeness with tweedle.

二层茶歇区临窗布置，呈现出较为稳定的状态，入座者更易感受到光线透过窗棂散落桌面的诗话景象。向南的尽头由横竖交织的排竹分割出茶座与电梯厅，并由向东延伸的排竹将用作洗手功能的饮马槽托举而上，颇有“四两拨千斤”式的巧力应和感，水源从顶面透过竹管顺流而下，饮马槽的沉重之势被瞬间削减。

二层包间区的入口被收纳在一个相对有压迫感体量内，“压迫”是为了更好地“释放”。在没有自然采光的现场条件下，取西边分割包间与公区的墙面凿壁借光，自然光线在通过茶歇区间后传递到包间内，虽没有斑驳感人的光线落入，却也不失温和透亮，白天被过滤后的光线在相对黑暗的空间内像一张开启光明的网。包间区过道内的墙面除了混合草茎的暖白腻子，亦有七百年历史的城墙砖陈设其中，行走其中具备时间的穿梭体验。包间内壁留白，取拙朴之姿态，给文人墨客留下足够的臆想与挥毫界面。

三楼全部设置为独立茶舍，依场地东西而立，交通中置，似林中小径，在南北进深三分有二处微微转折，借扭转之态，一个看似溪边草庐的建筑体离地而起，屋檐下探，竹窗由内而外撑起，似乎不论置身内外都有一探窗外究竟的愿望。

在狭长的过道中，为获得“静谧中探寻”的行走体验，并有效地将自然光线引入到一个并没有直接对外采光的封闭空间，曾几何时，回忆起某个很久以前的淳朴年代，门扇没有门套，没有踢脚，却在门扇上方有个邻里孩童打闹时，拴起房门依旧可以翻门而入，被唤作“亮子”的采光神器，可以解决在隔离中交换光线的问题，于是乎，存在于黑暗过道背光面的上部，并由竹篱叠加其中而形成的双层采光界面，充当了瞬时解放黑暗的勇士。而顶面转折处被雕塑化处理的局部搭建，正是在空间获得光的解放后所表现出的肆意姿态，有效地软化了相对硬朗的空间对接。包间依旧拙朴、留白。

入座，想起竹林七贤，想起耕读中的陶渊明，也许琴声起时，才是丰满。

TALKING CROPS OF GRAIN WITH WINE IN HAND

把酒话桑麻

Project Name | 项目名称
合肥市小灶王

Designer | 设计师
胡迪

Area | 项目面积
1500 ㎡

Main Materials | 主要材料
老木板洗白、水泥、细竹、夯土肌理漆、钢板、实木格栅等

Photographer | 摄影师
ingallery

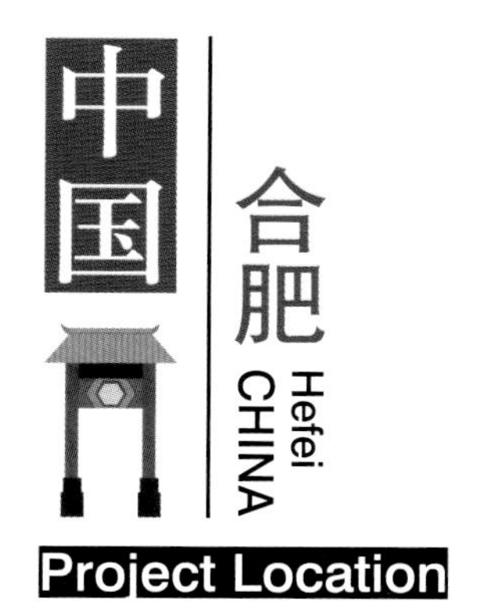

This case is located in the beautiful scenery of the East Dashu Mountain foothills of Hefei. The designer devotes himself to change the inherent impression of the traditional restaurant for regenerating the brand. The designer transforms Chinese natural rural village shape into inner building space. Through the clever unfolding layout and changing the scene per step, the designer strives to make modern urbanite return to nature and creates an idyllic poetic environment like "The village is surrounded by green wood; Blue mountains slant beyond the city wall. The window opened, we face field and ground; And cup in hand, we talk of crops of grain."

此案位于合肥风景秀美的大蜀山东麓，设计师致力于改变人们对传统餐厅的固有印象，让品牌焕发新生。设计师将中国自然田园村落的形态意向转化为内建筑空间，通过巧妙布局层层展开、移步换景，力求让现代都市人回归自然，构建出“绿树村边合、青山郭外斜，开轩面场圃、把酒话桑麻”的田园诗境。

果汁饮

The entrance of central nave uses the towering wooden fence which not only produces a sense of layer of regular space, but also becomes a symbol of accessing to the "the entrance of village". On the back of the wine bar, the old wooden boat swings duckweed in a leisurely calm like dreaming back into Huizhou. On the right side, a narrow alleys is formed between the house and walls built with steel and wood, and wandering in it, you will feel a distant mood. On the left, the semi-opening tile house alongside the water is arranged orderly with pavilion, surrounded by the bamboo and trees with water weaving through it, as if returning to the home garden.

The shadow of outdoor skylight is cut out by the iron lattice or bamboo fence exuding a unique charm, reflecting the traditional Chinese harmony philosophy and quietness and comfort which make people produce a leisurely feelings of returning to the rural landscape. Around the stairs, the steel modelling is embedded red glass with the flickering candle in it which makes the shadow gleaming like the shinning stars.

入口处的中厅用高耸的木制栅栏，让规矩的空间产生层次，同时也成为进入“村口”的象征。水吧背面的旧木船荡起浮萍，悠然自若，似若梦回徽州。中厅右侧，用钢结构与瓦木构建的房子与土墙之间形成狭长的窄巷，漫步其中，意境悠远。左侧区域的半敞开式临水瓦舍与亭子排列有序，四周竹林树木相映成趣，水系穿梭其间，仿佛回归乡土田园。

室外的天光通过铁艺花格或竹木栅栏，被切割出的光影，散发出独特的韵味，体现出中国传统“天人合一”的哲学思想，宁静安逸，让人们产生回归山水田园的悠然情怀。围绕楼梯的钢构造型嵌装红色玻璃，摇弋的烛火置于其间，让光影灵动起来，如星空般璀璨。

ENJOYING THE DELICACY

信步望花窗 雅趣享肴馔

Project Name | 项目名称
1718 公馆

Design Company | 设计公司
品悦公装

Designer | 设计师
曹会

Area | 项目面积
500 m²

The whole mansion is like Wu Guangzhong's painting, with the allocation of Suzhou lattice window and shadow, through the interweaving of point, line and plane, it performances an idyllic picture which not only has a rich oriental traditional charm but also features the characteristics of times, so that the guests find everything fresh and new.

The designer uses a large number of lattice window leak shadow to echo the elegance and fun of space and through the elegant poetry from inside to outside enhance the atmosphere to another level. Collocation of ancient wooden chairs, it is concise and orderly, and a large area of dark red rolls of chairs and seats make dining an elegant flavor.

The veranda of restaurant is extended, and the warm light shadow is dancing on the wood floor gracefully. The staggered partition walls are different from the restaurant environment. Using varying delicate types and changing patterns of lattice windows like "moon hole window" and "ornamental perforated window" as connection, its forms are various and lively. A single window can form a scene, while several windows can form a group of scenes. It can be said that it is not only blocked yet not crowed but also continued in the leak, which becomes the main focus of the restaurant. Strolling in a courtyard, the lattice windows are like natural frames to see different things and people from different perspectives. It is very interesting. If you can listen to a song of zither, and you will be intoxicated and enjoy a dining space like a painting.

整个公馆恰似吴冠中的水墨画，苏式花窗与光影配合，通过点、线、面的交织表现诗情画意，既富有东方传统意趣，又具时代特征，令观者耳目一新。

设计师运用了大量的花窗漏影，紧扣雅和趣，令空间从里到外透着典雅的诗意，将氛围提升到另一个层次。搭配古朴的木质桌椅，简洁有序，椅凳和卡座大面积的暗红色软包，让吃饭这一件事变得极富雅意。

餐厅游廊，幽径延伸，暖色的光影在木色系的地板上婀娜起舞，墙面隔断相互交错，营造出有别于餐厅大环境的一番天地。以“月洞”“漏窗”等各种精美的类型和形态多变的花窗作为连接，形式活泼多样，既能单窗自成一景，又能数窗形成组景，可谓隔而不堵，漏中又续，成为餐厅的最大亮点。闲庭信步中，一个个花窗就像天然的取景框，不同的角度，看不同的物与人，意趣盎然。如能听得一曲古筝，便更如痴如醉，享受像画一般的用餐空间。

A FUSION IN DIFFERENT STYLE

和而不同

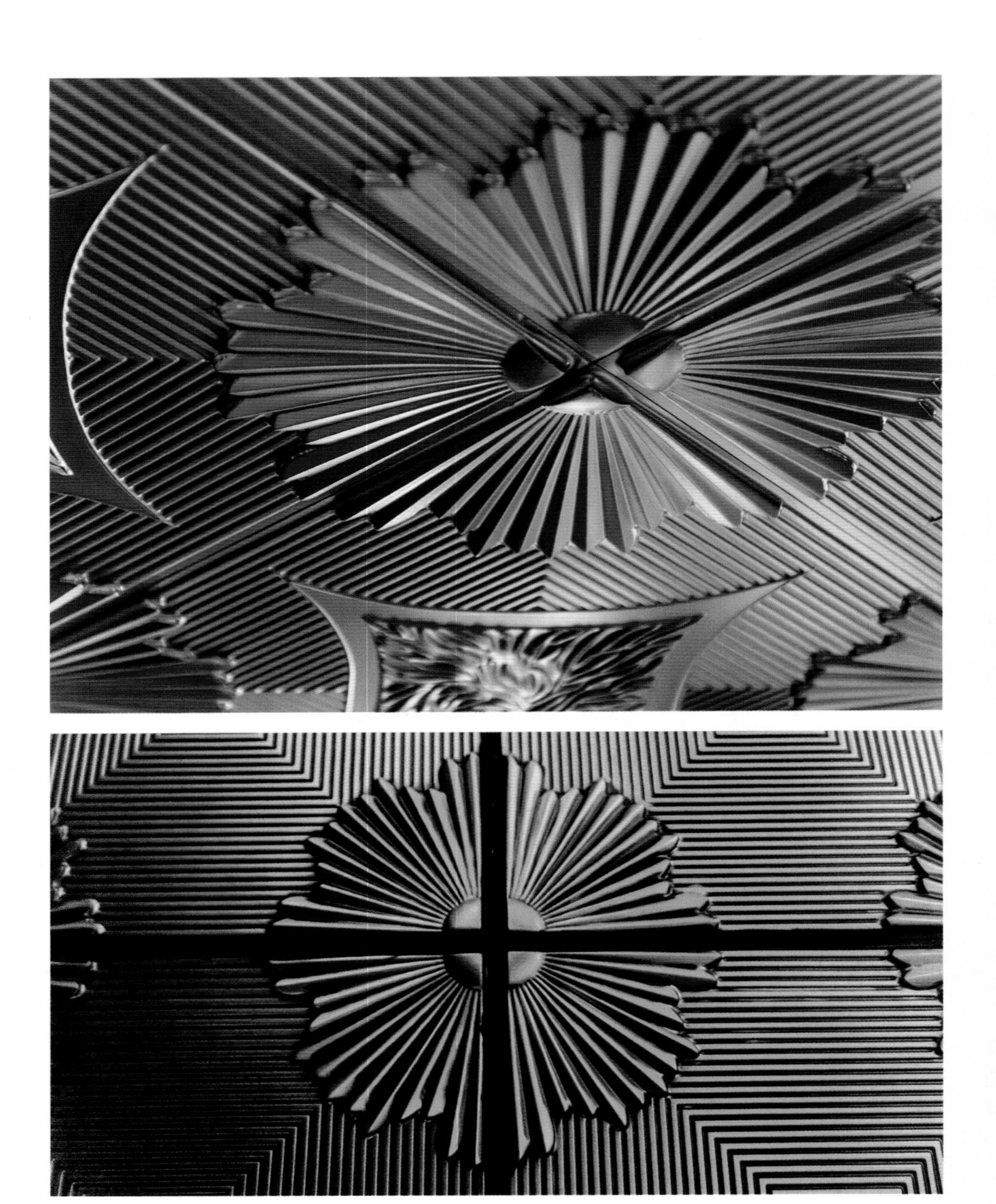

Project Name | 项目名称
BOTTEGA RISTORANTE& DIRTY LAUNDRY

Design Company | 设计公司
Einstein & Associates

Principal Designer | 主席设计师
Einstein & Associates

Area | 项目面积
600 m²

Bottega Ristorante & Dirty Laundry is one of European concept restaurant and lounge in Jakarta's central business district; SCBD - South Jakarta, Indonesia. The concept for this specific restaurant is 'La Dolce Vita'; meaning the good life, a life of pleasure and simple luxury. It is about enjoying the good things in life and indulging in the things you love. Bottega Ristorante & Dirty Laundry use its 1920s Art Deco style to bring new atmosphere with modern glam. The art deco style itself represents luxury, glamour, and faith in social and technological progress. It combines modernist styles, geometric forms with fine craftsmanship and rich materials.

The Art Deco style is adopted in Bottega Ristorante & Dirty Laundry and can be seen in the mixes of materials like a gold mosaic, brass and metal plate at the ceiling tiles put into modernistic forms, black steel on the frontage of the building to give an idea of solidity. All the design elements of Bottega Ristorante SCBD and Dirty Laundry represents the art deco touch with a more sleek form of the style; it featured curving forms and smooth, polished surfaces.

BOTTEGA

BOTTEGA

BOTTEGA

The Art deco visual style is used to be the glue that holds local materials and modern tropical design elements together and symbolizes the contemporary European style, considering that Art Deco is a style known for its elegance and chic. Strong lines, bright colors, glossy materials and geometrical shapes are not a strange scene in this restaurant. All the elements simply scream to be noticed by people who walk past it.

This restaurant is divided into two areas: the dining area that is Bottega Ristorante and the lounge area introduced as Dirty Laundry.

The dining area consists of alfresco area, semi-outdoor area and indoor area. Entering the premises, guests are welcomed by the alfresco area with beautiful vertical garden on the back of a long bench. The main point of the alfresco dining area is the brass reception table. Continuing to the main entrance, the guests enter the foyer area dominated with oak wood panels wall and ceiling. This corridor is the transition area between the alfresco, semi-outdoor and dining area. On the right, is the semi-outdoor area, where guests can see and feel like they're in the outdoor area, but still in the indoor area. This area resembles the previous design of Bottega Ristorante in the old location, with the long and round bench, peacock mosaic table and mirrors to mimic the windows.

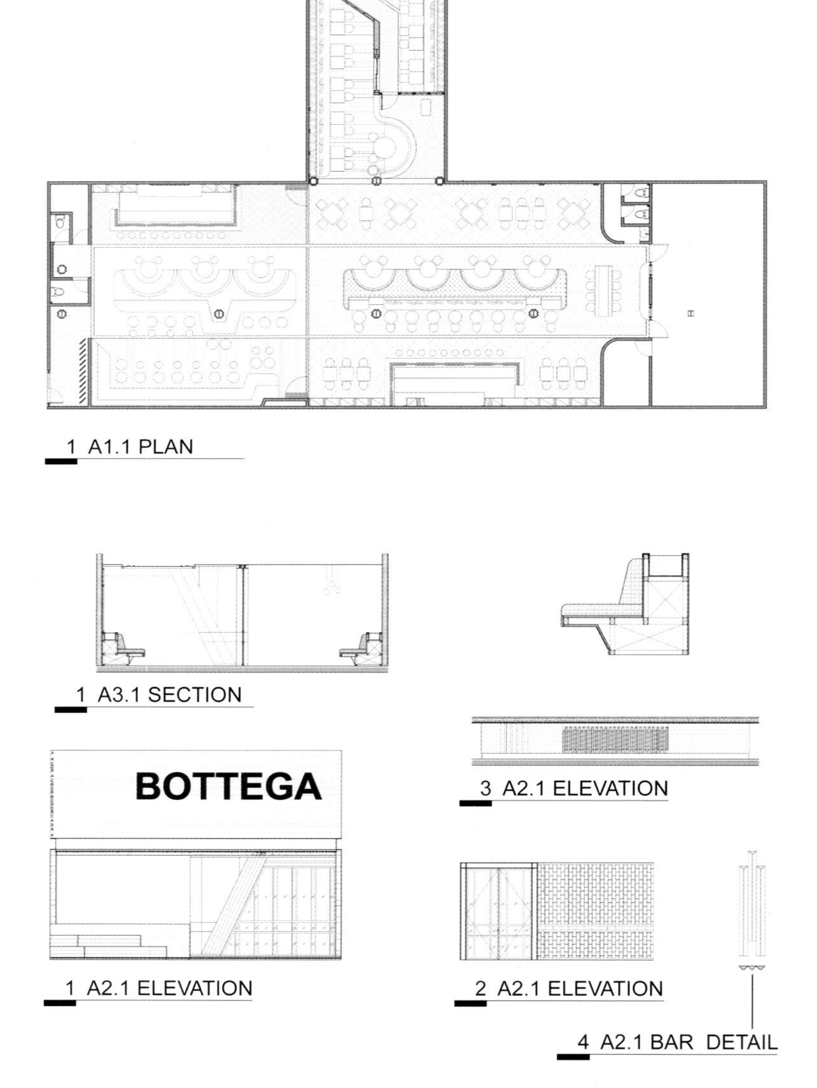

Bottega Ristorante & Dirty Laundy位于印度尼西亚，雅加达南部中央商务区，是一家欧洲概念餐厅和休息室。这家特殊餐厅的理念是“La Dolce Vita”，寓意着好的生活，一种生活的乐趣和简单的奢侈。它是关于享受生活中的美好事物，沉迷于你所爱之事。Bottega Ristorante & Dirty Laundry运用了20世纪20年代的装饰风格，营造了一股时尚迷人的新气息。这种装饰艺术风格本身代表着奢华、魅力和对社会技术进步的信心。它集现代主义风格的几何造型、精细工艺和丰富材料于一体。

Bottega Ristorante & Dirty Laundry沿用了这种装饰艺术风格。天花板瓷砖采用了金箔、黄铜和金属板等现代元素；餐厅的正面用黑钢来体现其坚实牢固的理念。南中央商务区的Bottega Ristorante与Dirty Laundry的所有设计元素呈现出一种光滑触感的装饰风格，带有弧形与光滑、抛光的表面。

该案例提取当地的材料并结合现代热带设计元素来象征着当代欧洲风格。这种装饰艺术以其优雅和别致著称。线条强烈、色彩鲜艳、材质光亮和各种几何造型的使用都不足为奇，所有的元素都只为了吸引路过的人们。

该餐厅分为两个区域：Bottega Ristorante就餐区和Dirty Laundry休息室。

The ceiling of the dining area uses the 1920 art deco brass ceiling tiles, and the floor uses herringbone patterned oak hardwood floor combines with patterned hexagonal tiles. The Dining area itself is divided into three areas: side dining, VIP dining and bar dining area. The long sofa bench at the center of the dining area separates the side dining and bar dining area. The center point of the dining area is the floral mosaic on the back of the bar. On the other side, lies the VIP dining, standing exclusively separated on its corner right between two half-circle ivory travertine wall and the iconic peacock mosaic of Bottega Ristorante at the middle. The bar and the wine shelves become the background of the whole dining area. The black marble top table combined with brass rivets looks grand even from the front of the restaurant. On the right, there is a gold mirror partition that separates Bottega Ristorante and Dirty Laundry. The two areas can be opened as one if the partition is fully folded.

Dirty Laundry is a lounge and bar in Bottega Ristorante. The interior concept of dirty laundry still has a connection in the art deco style just like the restaurant. Dirty laundry also uses 1920 art deco ceiling tiles. The floor uses black and white hexagonal mosaic tiles, mixed with ebony hardwood floor. The walls are accented with patterned ceramic tiles with a copper plate. The sofa's color is mint green to light up the otherwise rather dark atmosphere.

Even though the two can stand on their own, with different design, the designer connects Bottega Ristorante and Dirty Laundry in several aspects of the design. The symmetry designs between Dirty Laundry and Bottega Ristorante's ceiling connects two areas. When the restaurant and lounge opens, the door partition slides into one side and the two areas mixed like they are one right from the start. Linearly, sofa booths in both areas and the floor patterns becomes one; they can only be differentiated by the darker colors in one of the area.

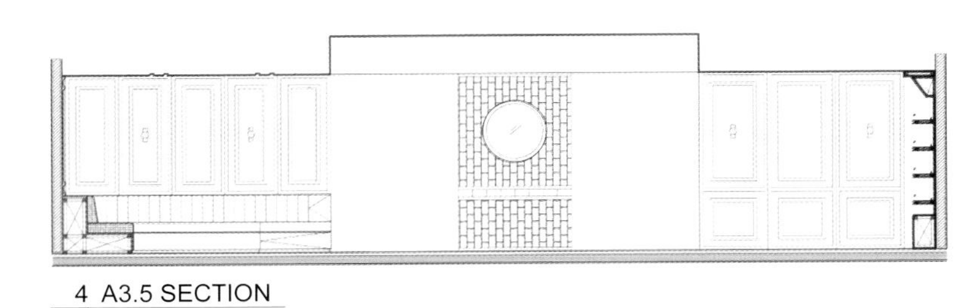

4 A3.5 SECTION

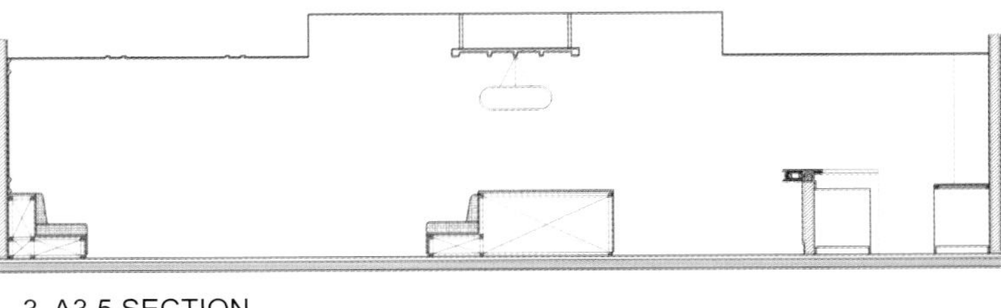

3 A3.5 SECTION

就餐区由露天区、半室外区与室内区组成。进入餐厅场所，顾客首先看到的是位于美丽的垂直花园长椅后的露天区。黄铜接待台是露天就餐区的亮点。来到主入口，顾客们走进橡木木墙板和天花板的门厅区。右边是半室外区，在那里顾客可以坐在室内看到和感受到室外的风景。这个区域与Bottega Ristorante旧址的设计相似，有着长圆形的长凳、孔雀马赛克的桌子和窗户形状的镜子。

就餐区的天花板使用20世纪20年代装饰风格的镀铜吊顶板，地板采用人字形橡木硬木地板搭配有图案的六角形瓷砖。就餐区划分为三个区域：大厅区、贵宾区和吧台区。就餐区的沙发长凳分隔开了大厅区和吧台区。吧台后面的花朵图案是就餐区的亮点。另外，贵宾区被两个半圆象牙洞石壁完全分离开来，其中间具有餐厅标志性的孔雀马赛克瓷砖。吧台和酒柜作为整个就餐区的背景，黑色大理石的桌面搭配黄铜铆钉，即使从餐厅前面看来也很是壮观。右边的金色镜子将餐厅与酒吧隔离开来。如果把镜子完全折叠起来，餐厅与酒吧即可合二为一。

Dirty Laundry是Bottega Ristorante的休息室和酒吧。酒吧的设计理念延续了餐厅的装饰风格。Dirty Laundry同样采用了20世纪20年代装饰风格的吊顶板。地板是用黑白六角形的马赛克瓷砖来搭配乌木硬木地板。墙面突出强调了带有图案的铜板陶质瓷砖。薄荷绿的沙发照亮了原本相当黑暗的气氛。

即使这两个区域可以坚持自己与众不同的设计，但是设计师通过几个设计元素就将Bottega Ristorante和Dirty Laundry联系起来。Dirty Laundry和Bottega Ristorante对称的天花板就将这两个区域连接起来。当餐厅和休息室同时打开时，门的隔板推向一边，两个区域即可融为一体。沙发与地板上的图案也融为一体。人们只能通过颜色的深浅来区分两个区域。

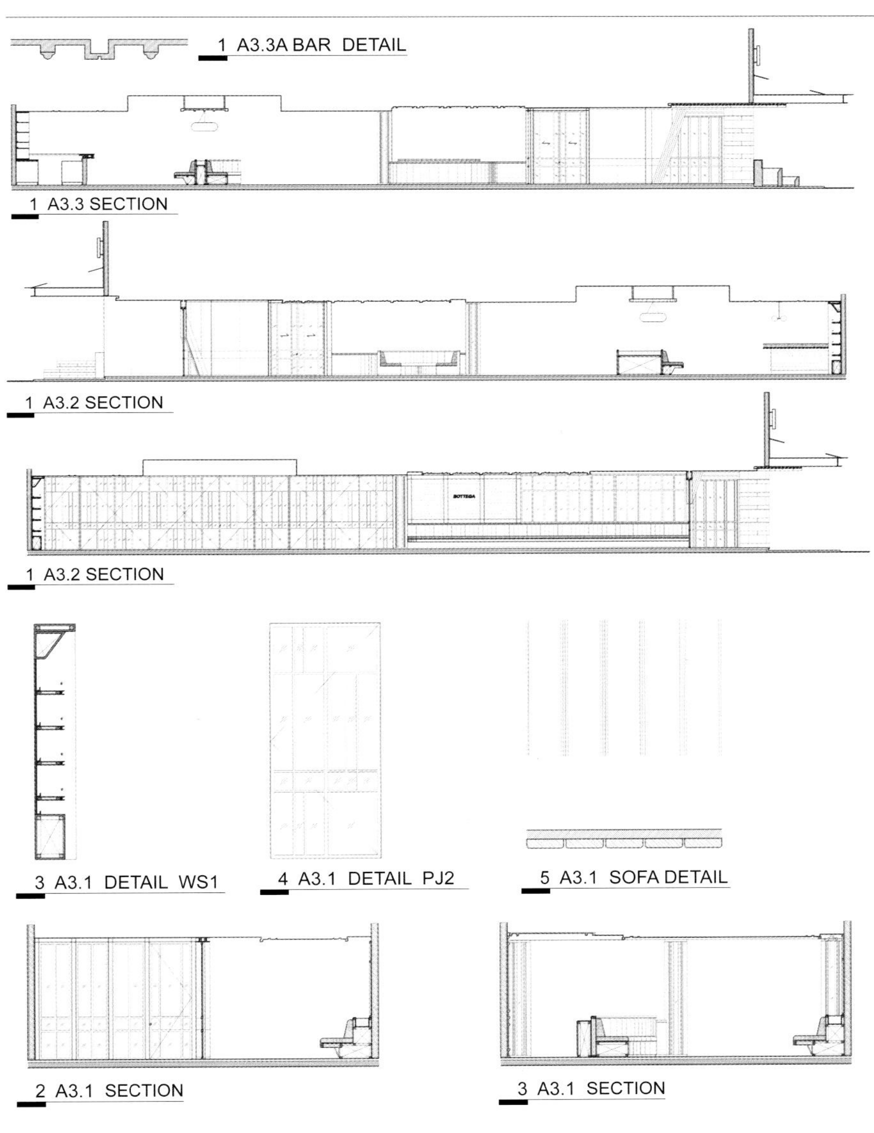

DISLOCATED SPACE-TIME WITH TIMES GOING

错位时空 岁月流转

Project Name | 项目名称
Y2.space 元色餐厅

Design Company | 设计公司
重庆燚筑纵合室内设计有限公司

Designer | 设计师
刘攀

Furnishings Designer | 陈设设计
王晓蒙

Participant Designers | 参与设计
徐再攀、陈杨、邓义川

Area | 项目面积
500 m²

Main Materials | 主要材料
黑钢、镀锌板、橡木地板、铁管、仿铜板等

Photographer | 摄影师
张骑麟

As fashion becomes a kind of art and dining becomes a visual contest, it's necessary to create a fashion yet personality high-end restaurant so that the Primary Color restaurant tries to present a unique visual delicacy feast to people.

In this case, the perfect integration of design and space is amazing. The immovable column and flexible runway form a strong visual contrast, resembling the scattered meteorolite in the universe. Polarized vision always can bring people a sense of avant-garde fashion.

The extended curve is the combination of light and shadow, black and white, and the curved runway seems to be endowed with a great extensibility in the light figure of the model, like the universe continues alternately day and night. In the space, the faint shadows and interspersed lines are elegant and scattered, as if this is not a fashionable high-end restaurant, but a magical place for people to enjoy the day and night circulation of the universe.

Scattered flush light panels are interspersed the whole space to enhance the dimensional sense of space, resembling the stars of the universe, bright and shine but not dazzling. From the perspective of kaleidoscope, it fulfils with colorful changes as clouds in the meteorite shape and flying butterfly, which mingles with black and white, you can embrace all of these natural extreme beauty when you are dining.

Throughout the entire space, the designer integrates the original spatial structure cleverly and uses invisible, ethereal elements to change the three-dimensional living space into a unique art exhibition. As the designer said, “Fashion, art and life eventually cannot last without the past time.” This stereoscopic restaurant like the universe resembles a universal show which presents a wonder of art and life.

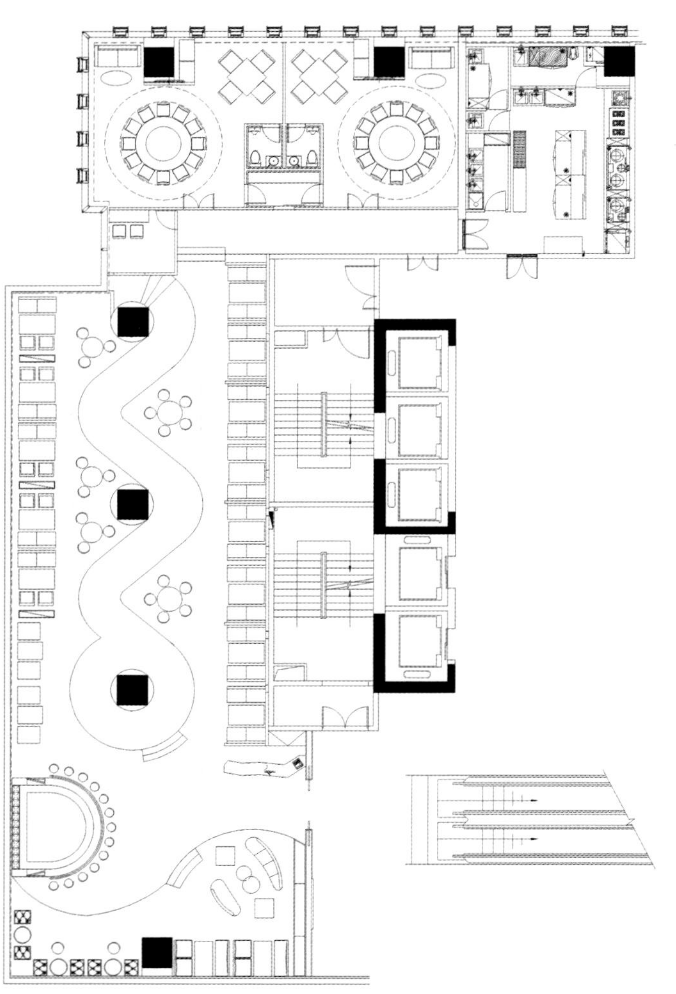

当时尚成为一种艺术，当用餐早已成为视觉较量，打造一个时尚与个性融为一体的高级餐厅势在必行，元色餐厅力图呈现给众人独特的视觉美食盛宴。

本案设计与空间本身浑然天成，两者的高度融合让人惊叹。柱子的不可移动性与灵动流畅的T台在视觉上强烈碰撞，正如宇宙间陨石的散落，两级分化的视觉总能带给人前卫的时尚感。

延伸的曲线是光与影、黑与白的汇聚，曲型的T台在模特们轻盈的身影下似乎被赋予了极大的延展性，回环往复，如宇宙日夜交替般延续。空间里阴影恍惚、线条穿插、飘逸错落，仿佛这不是一个时尚高端的餐厅，而是一个众人一起欣赏宇宙昼夜流转的神奇处所。

零星散布的筒灯穿插于整个空间，如宇宙中的星，明亮闪烁却不刺眼，为立体感的空间增色。万花筒的视角里充满着炫彩的变幻，正如空间里的陨石状浮云、飞舞灵动的蝴蝶，透明与黑白交织，用餐的瞬间收揽了自然的极致美。

纵观整个空间，设计将空间原态结构巧妙融合，运用隐形的飘逸元素，让立体化的生活空间变成了一个独特的艺术展。正如设计师所说，“时尚、艺术、生活终究离不开岁月的流转”，这个拥有宇宙般立体感的餐厅，如错位的时空秀场，充满灵气，让人不得不赞叹艺术与生活的奇妙。

MODERN ATMOSPHERE, FRENCH ROMANCE

现代风情 法式浪漫

Project Name | 项目名称
Mr & Mrs Bund

Design Company | 设计公司
Kokaistudios

Chief Architects | 设计责任人
Andrea Destefanis, Filippo Gabbiani

Design Manager | 设计经理
Kasia Gorecka

Design Team | 设计团队
付炼中、王思昀、刘永泰、Suju Kim、孙文娟、顾庆龙

Area | 项目面积
1300 ㎡

Photographer | 摄影师
夏宇

In Chef Pairet's vision, Mr & Mrs Bund is a casual dining space that is free from the confines and norms of traditional fine dining. Space should create a sensation, not only through gastronomic wonders stimulate diners but also visually keeps diners' eyes moving by the environment.

Aware of the impact of light on human psychology, designers have designed the lighting system of Mr & Mrs Bund to create two distinct moods. A soft and elegant space filled with gentle sunlight during the day, and the restaurant becomes much more dynamic, romantic and posh at night. The duality of character invokes different emotions among diners, and it makes Mr & Mrs Bund one of the rare places where diners would like to visit day and night.

在主厨Pairet的愿景中，Mr & Mrs Bund应是休闲放松的用餐场所，不受传统美食的限制与束缚。此空间应创造出全方位的感官体验，除了美食给味蕾带来的刺激，更用美妙的环境牢牢抓住食客的视觉。

基于光线对人心理的影响，设计团队为Mr & Mrs Bund设计的灯光系统创造了两种截然不同的情绪：充盈自然光线的柔和而优雅的日间；浪漫精致中不乏活力的夜间。二重性调动了就餐者不同的情绪，这让Mr & Mrs Bund成为来宾们不论白天或晚上都欣然到访的独特餐厅。

Throughout the restaurant, designers utilize classical French motifs and a palette of materials typical of French traditions, in an intelligent and irreverent way that creates an adaptable environment suitable for an elegant dining experience or bigger, crowd-drawing parties. The original building elements such as windows and pillar decorations have been preserved and treated with a gray unfinished plaster, and stereoscopic wallpaper highlights the building's classic interiors in the past century. The ceiling has been treated with a patchwork of custom designed hexagonal tiles to create a stratified unfinished effect, whilst the parquet timber floor is laid with an unexpected composition of colors and a rustic polishing finish.

All the newly added partitions have been graphically treated with classic French boiserie photography applied. They are infused with the traditional technique of trompe-l'oeil in an extreme way, enriching the space with a strong identity and opening up the walls with an interactive game of illusion. Decorations are also placed on walls to create a stronger multi-dimensional visual experience.

Seating is a particular highlight of Mr & Mrs Bund. The furniture reuses classic typologies of the French tradition with innovation. For instance, typical wood chairs have rough and unfinished frames, with fabrics in different hues of denim and seat backs decorated with corsetry and vest details.

The armchairs of banquet table are fabricated from handmade iron bars, corresponding with the theme of early industrialization. Bar and lounge stools are treated with graffiti and abstract paint motifs, whilst dining sofas are finished with textile commonly used for potato jute bags. Most seating is in a neutral color, except for the traditional egg chairs behind the reception and the long banquet table chairs, which highlight the central dining area with a dash of red and green.

设计师将经典的法式图案纹样及典型法式面板材料运用于整个餐厅，以一种智慧和幽默的方式创造出优雅的就餐体验和各种规模人群的可适应性。原先建筑中的元素诸如窗户和柱子的装饰被保留下来，处理成灰色未完成感。立体墙纸彰显出20世纪经典的室内风格。天花由定制的六角形瓷砖拼接而成，创造出分层的未完成效果，同时木地板采用了不经意的组合和朴素抛光面的处理方式。

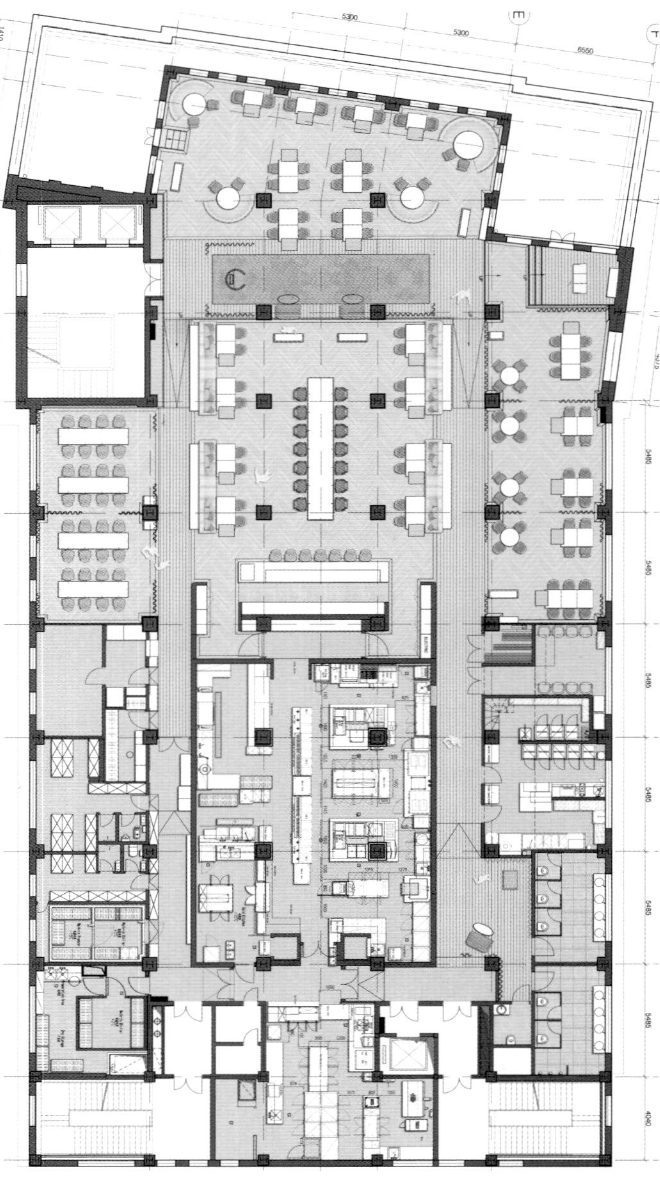

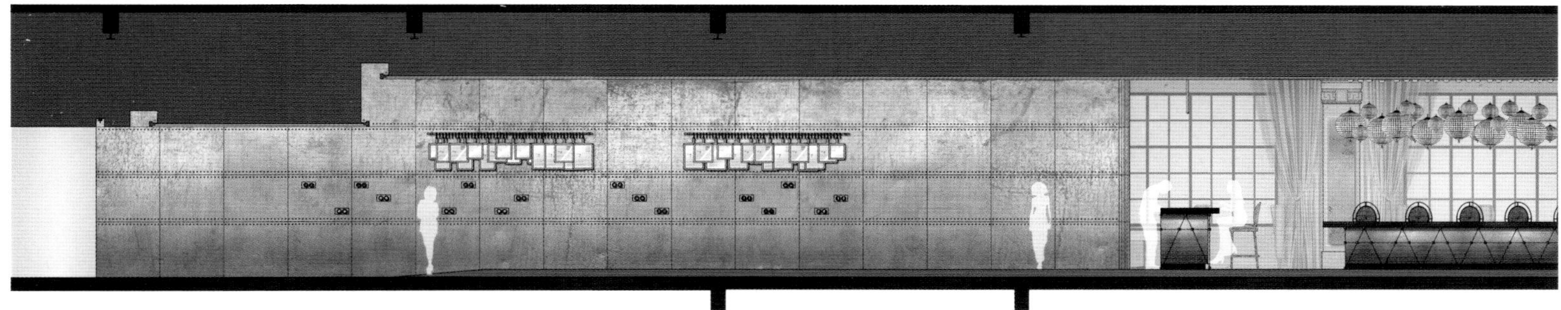

所有新增的隔断都被图示化地处理为法国经典的细木护壁板图像，传统的技艺中加入了视觉陷阱，强烈的身份标识丰富了整体空间，互动游艺的错觉使墙面更显开放。墙面的装饰创造了强烈的多维度视觉体验。

座位是Mr & Mrs Bund的亮点，座椅沿用了法式经典风格而又有创新。例如，典型法式木椅，椅背木框表面粗糙呈现未完成感，配以不同色调的纹粗棉布布料。

宴会桌的靠背椅框架由手工铁条制成，呼应早期工业化的主题。吧台和休闲区的座凳被做成涂鸦和抽象画的主题，同时就餐的沙发用做土豆麻袋的织物制成。大部分座椅是中性色，除了在前台后方传统的鸡蛋椅和宴会桌配椅，它们明亮的红绿色点亮了中央就餐区域。

FULL OF FLOWERS
A BLUE FAIRY TALE
鲜花铺酒·蓝色童话

Project Name | 项目名称
Café Oct.22

Design Company | 设计公司
北京海岸设计

Designer | 设计师
郭准

Area | 项目面积
857 ㎡

Main Materials | 主要材料
玻璃、混凝土、钢、木、砖、石等

Everyone has a dream of a fairy tale, dreams of becoming the prince or princess of the story, and living a happy life in the beautiful, luxurious castle. In the space, the designer uses the blue color as the main tone, upholds the concept of "return to origin", abandons the complicated texture and decoration in dealing with details and integrating the French pillars, carved, crystal chandeliers and candlesticks into it to create a beautiful and noble atmosphere, making customers seem to be in a luxurious castle full of fairy tale love.

每个人心中都有一个童话梦，梦想自己可以成为故事中的王子或公主，在美丽奢华的城堡中过着幸福快乐的生活。在空间中，设计师以蓝色为主色调，秉持归本主义理念，在细节处理上摒弃过于繁杂的肌理和装饰，将法式廊柱、雕花、水晶吊灯、蜡台等元素融入其中，营造出唯美、高贵的氛围，让来客仿佛置身于充满童话爱情的奢华古堡里。

The plants hanging on the French pillars flexibly "separate" the space, with the embellishment of several bouquets of lilies which people can feel a sense of romance and elegance. The white carved tables and chairs with fashionable appearance and graceful lines outline a romantic and elegant style. Under the reflection of light and lamps, space is warm and comfortable. The crystal chandelier down from the ceiling shows its precious and extraordinary elegance. Iron windows with rolling grass pattern can broaden the vision of space, integrating the outdoor landscape with interior decorations perfectly.

The height-raised area uses a large crystal chandelier as decoration, and the stairs design gives the static space a sense of dynamic and sublimates visual environment of the interior. The dressing room and the restroom are based on blue background, and the long wooden tables, bright flowers and exquisitely carved mirror add a bright finishing color for interior space and a sense of bright spring.

法式廊柱结合悬挂而下的绿植，灵活地“隔”断了空间，偶有几束百合花点缀，浪漫清新之感随即扑面而来。米白色雕花桌椅以时尚的外观造型、优美的线条轮廓，勾勒出浪漫典雅的风格。光与灯的映衬下，空间温馨而舒适。水晶吊灯自天花优雅垂落，尽显矜贵不凡的优雅。卷草纹铁艺窗打开空间视野，将室外风景与室内装饰完美融合。

挑空区以大型水晶吊灯为装饰，盘旋而上的楼梯设计赋予了静态空间动态感，升华室内的视觉环境。化妆间、卫生间以蓝色布景为主，长条木桌、鲜艳花卉、精美雕花镜为空间增添了亮丽的内饰色彩，给人以明媚的春光感。

Cafe Oct.22

THE ROMANTIC OLD SHANGHAI

海派·浪漫老上海

广州
Guangzhou
CHINA

Project Location

Project Name | 项目名称
锦秀里火锅

Design Company | 设计公司
广州泽辰设计

Area | 项目面积
320 ㎡

Photographer | 摄影师
赵彬

The CBD Zhujiang New Town and the tree-lined Lingnan dwellings with gray brick form a conflict of bustling fashion and traditional quiet and bring a visual surprise. Jin Xiu Li is located in Guangzhou Liede rive edge, designers want to retain and continue the surprises from the conflict of environment in the design of Jin Xiu Li so that designers use a retro green color as the main tone blending a three-storey building with retro modern elements, like a piece of exquisite jade inlaid in the Canton Tower, wipe out the spark in the elegant south of the Five Ridges culture and exquisite female perspective.

高楼大厦林立的CBD珠江新城，青石灰砖、绿树成荫如世外桃源的岭南民居，形成了繁华时尚和传统幽静的冲突，带来了视觉上的惊喜。锦秀里正坐落于广州猎德河涌边，设计师希望能够保留和延续环境的冲突带来的惊喜，因而一座以复古绿色为基调，糅合了复古摩登元素的三层小楼，最终犹如一块精致的翡翠，镶嵌在广州塔下，在灵动的岭南水乡文化与细腻的女性视角里擦出雅致的火花。

锦秀里火锅

锦秀里火锅

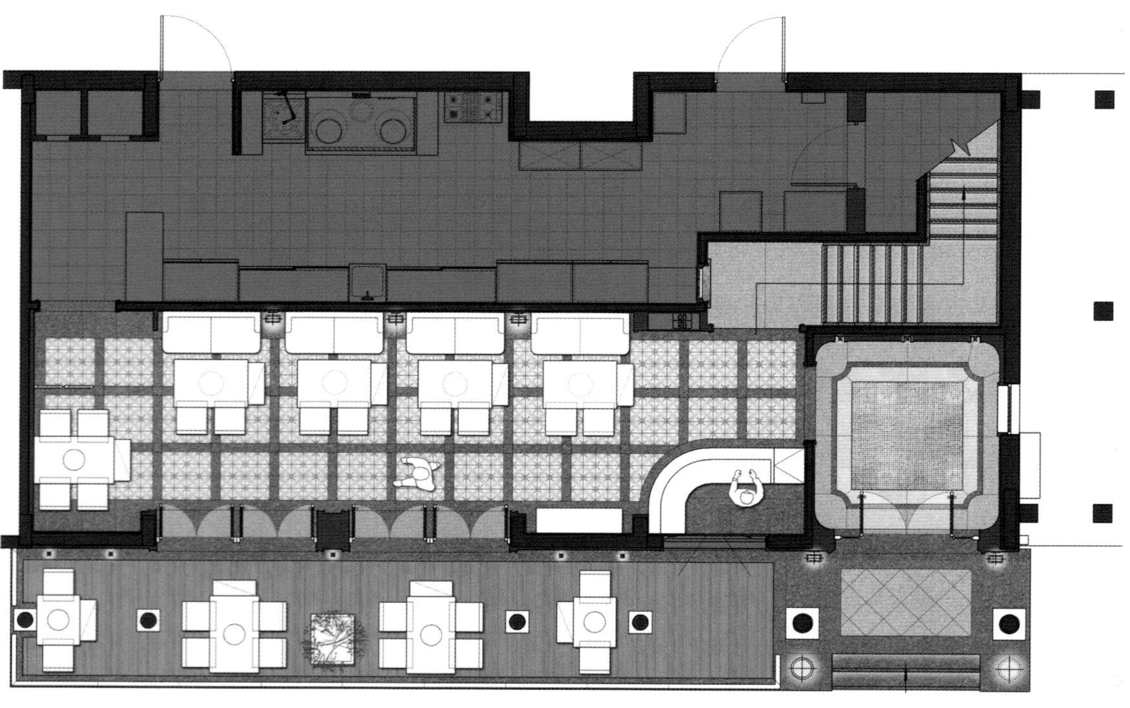

Entering in the Jin Xiu Li, you can see a bold blue and green color, like the elegant emerald, or the remarkable peacock blue with little yellow jumping out of dull. Tiles, velvet curtain, carved windows and other unique decorative items are filled with Old Shanghai style. The eye-catching calendar girl almost is one painting in one step. The embellishment of the wall is a rich color art hand-painted paintings, and the soft light penetrating through the ceiling reflecting the dark wood rattan table, lake blue cloth cane chair and the herringbone parquet floor exude an extravagant flavor in a comfortable atmosphere and heighten an old Shanghai retro flavor and romance. The conflict between tradition and fashion, and the wharf culture behind the hotpot and the female theme demanded by the owner create a new and elegant dining atmosphere.

进入锦秀里，迎面而来的是色调大胆的蓝绿色，似典雅祖母绿，又似张扬孔雀蓝，偶尔的些许黄跳脱出沉闷。花砖、绒幕帘、雕花窗棂等别致的装饰性物件洋溢着十里洋场的海派风格。吸引人眼球的挂历女郎，几乎一步一幅。墙身以富有色彩艺术的手绘挂画做点缀，天花渗透柔柔灯光，映照着深木色的仿藤餐桌、湖蓝色布艺藤椅、人字拼花的木地板，为一片惬意的氛围增添贵气，烘托出老上海的复古味道与浪漫。传统与时尚的冲突、火锅背后的码头文化与经营方要求的女性主题之间的冲突，产生了一种全新的、优雅的餐饮氛围。

JIN XIU LI

JIN XIU LI

宋锦

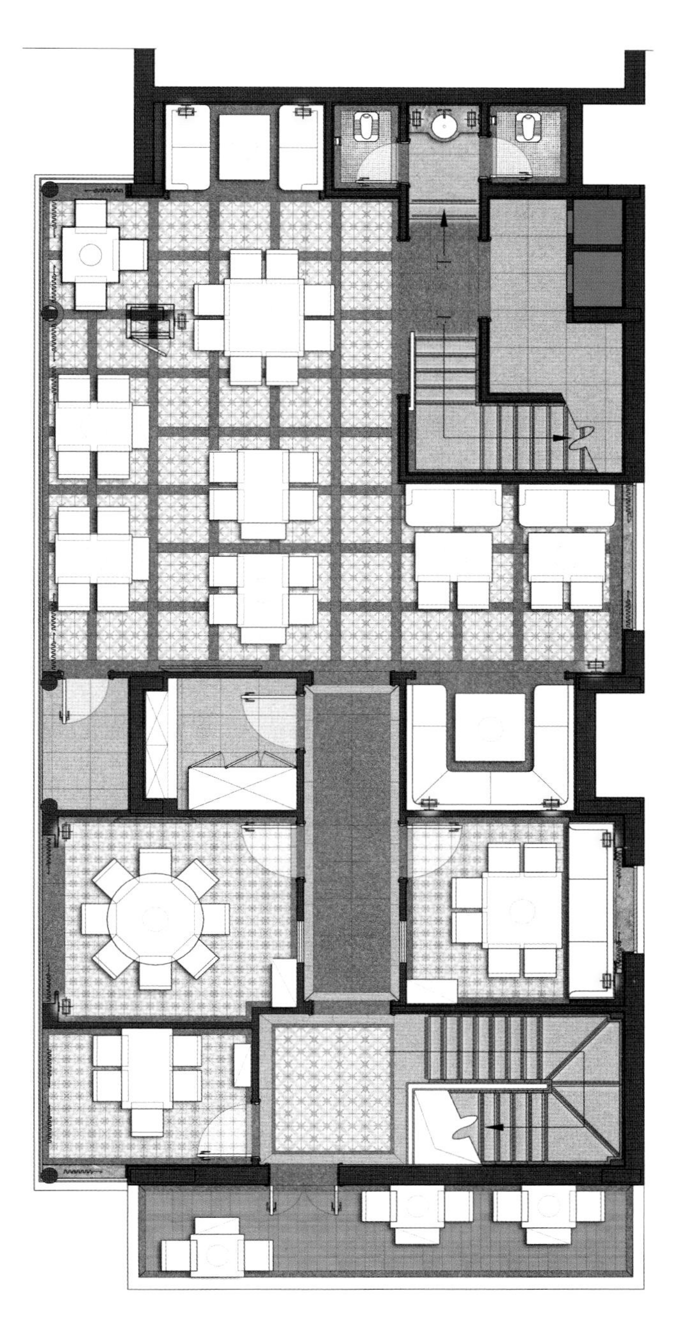

THE DELICACY AND PARADISE WITHOUT BOUNDARIES

无界美食·乐园

中国 北京 Beijing CHINA

Project Location

Project Name | 项目名称
Jason 无界美食

Design Company | 设计公司
寸 DESIGN

Designer | 设计师
崔树

Area | 项目面积
360 ㎡

"Restaurant could be unlike a restaurant, and the booths couldn't have to be placed in the corner. Rather than a simply dinning place, it should be more like a place suitable to have a party with close friends." There are no boundaries between cuisine and wine, also the design is without boundaries.

This is a restaurant transformed from a cool select shop, a redefinition of space by the designer. The great changes are preserved by the designer that is from calmness and independence to bustle and fusion, dilapidated American style to gorgeous European style. The designer transforms the original vertical ladder with the wall into a revolving staircase in the center of the space, retains the rust feeling of the metal with the industrial elements to make the space more modernized. At the same time, the metal revolving staircase makes the indoor structures more impressive, strengthens the form feeling on the first floor, and simultaneously makes room for bar.

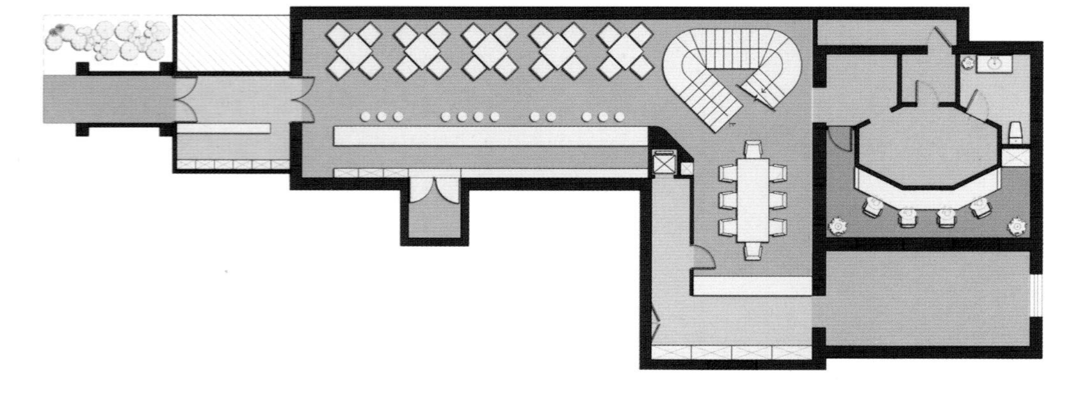

The designer chooses hexagonal floor tiles to pave the ground for the first floor, and the wall and plants in the ceiling also adopt the hexagonal elements. The interspersed curved staircase in space adds a new DNA for the space, resolving the stiff feeling because of a number of linear types and "L" shapes in space to make the space more abundant and more personalized.

Next to the staircase on the second floor, there is the restoring murals, *The Garden of Earthly Delights*, which its original is an oil painting. The designer processes it into European old mural texture which uses the kaolin as the printed baseplate, and then prints the picture on it, so that retains the original form of cob bricks. The wonderful scenes of life in the paintings, like the changing scenes and functions in the restaurant, are matching with the restaurant perfectly. Without any unmovable booths, this restaurant erases the boundaries between the Chinese-Western food and boss and guests, and makes space for flexibility and possibility of permutation and combination. Even if there are endless parties, it will not affect other customers' experiences. While the spiral staircase makes this whole restaurant a segmentation and regional store. Creating an atmosphere in the design can strengthen the interaction between people and people. Here, people do not interfere each other, but also can contact each other, independent yet bustling.

“餐厅也可以不像餐厅，卡座也不一定就放在某处角落，比起单纯的进餐场所，它应该更像一个不在家中，但适合和亲密朋友聚会的地方。”菜品、酒品无界，设计亦无界。

这是由高冷的买手店转型而来的餐厅，是设计师对空间的重新定义。从冷静独立，到热闹融合；从破败美式，到略带欧洲风的华丽，变化巨大，设计师将一处处元素保留了下来。将原本贴着墙的直梯改为在空间中央的旋转楼梯，保留金属锈迹斑斑的感觉，加入工业化的元素，让带有时间感的空间更加现代化。同时，金属的旋转楼梯让室内结构更加可观，强化一楼的形式感的同时，也给吧台让出了位置。

设计师选用六边形地砖铺设一层地面，一层的墙面和天花板的绿植格亦采用六边形元素，加上曲线的楼梯在空间里穿插，为空间添加一个新的DNA，化解空间里过多的直线形、“L”形带来的僵硬感，使空间更丰富、更有性格。

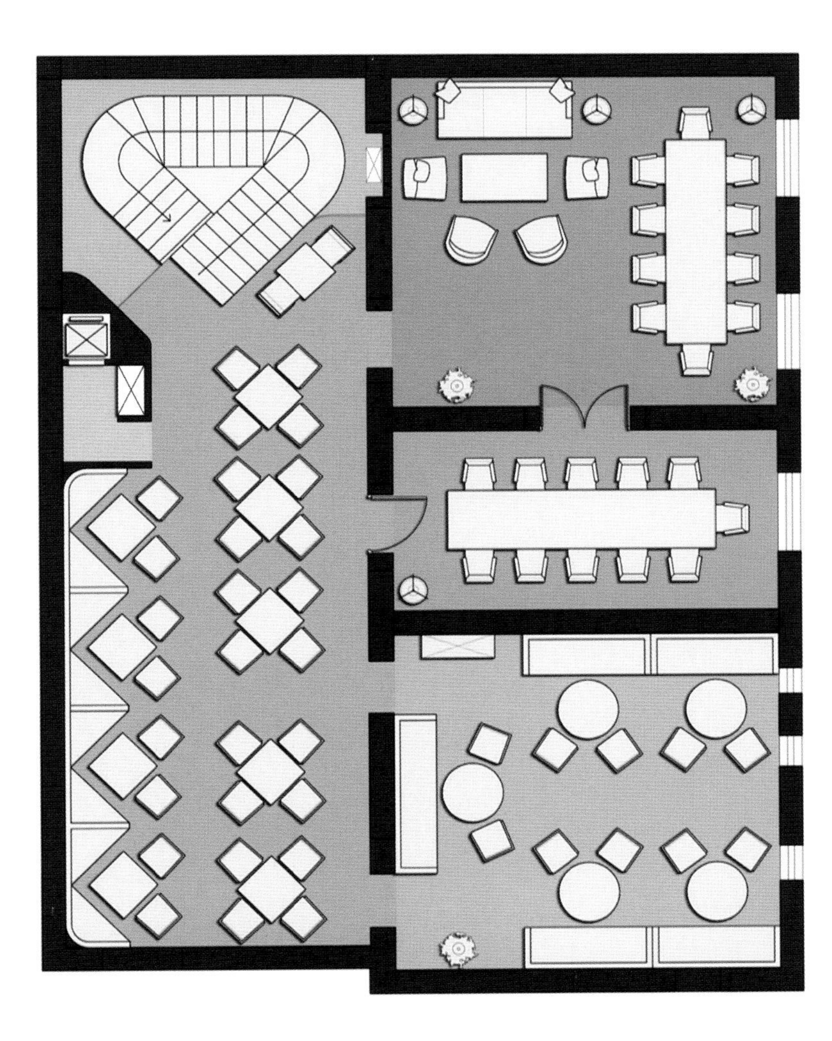

二层楼梯旁边的复原壁画《人间乐园》的原作是一幅油画，设计师把它处理成欧洲老壁画的质感，把高岭土做成可以打印的底板，再把画面打印在上面，保留了原本土砖一块一块的形式。画里各种奇妙的生活场景，一如餐厅里不断变化的场景和功能，与餐厅十分搭调。不设置任何不可挪动的卡座，擦除中、西餐与老板、客人的界线，为空间留出更多灵活机动以及排列组合的可能，即使派对不断，也不会影响其他客人的体验。而螺旋形的楼梯，又把一个整体的店变成切分的、有区域的店，在设计上营造氛围，让人和人之间的交互性得到强化。在这里，人和人既不互相打扰，也彼此联系，独立却不失热闹。

JASON
無界美食
Boundaries

MOUNTAIN SHAPE

山非山，水非水

Project Name ｜ 项目名称
牛牛西厨品牌连锁餐馆（乐从店）

Design Company ｜ 设计公司
硕瀚创研

Soft Decoration ｜ 软装陈设
东西无印

Principle Designer ｜ 主持设计
杨铭斌

Area ｜ 项目面积
530 ㎡

Main Materials ｜ 主要材料
木饰面、水泥艺术涂料、工艺瓷砖、天然麻石等

There is a large glass floor curtain in the building where the project locates, and the transparent building provides more possibilities for the presentation of the interior space. In the design, space is divided into two spaces of different heights. First of all, the front part at the entrance is the visual highlight of the entire space and the external environment to attract each other. Taking advantage of the building's natural advantages, the front part of the building is designed to be a high-ceilinged space to echo with the restaurant's "garden" theme. A coarse material rich in natural texture—rope is selected to design a set of "mountain shape" art installation with line permutation and combination at the highly ceilinged space and the upper part of the ladder. Besides, bulb lights are interspersed in the "mountain shape" just like the stars on the winding mountains. In the design, the coarse materials and refined design make the rope have far more valuable than its own.

Indoors, the lowest point except the pipeline is only about 5 meters, which is an awkward height. In the design, the designer needs to consider adding additional mezzanine to meet the owner's business needs. Therefore, the latter part of the space is designed into a two-layer dining area. While guaranteeing the height of the lowest point is at 2.4 meters, the mirror stainless steel material is used as a layer of the ceiling tiles. Because the stainless-steel mirror surface is rough and uneven, the ceiling has a dreamy effect like the water wave, which reduces the sense of oppression caused by the insufficient height of the layer.

In the design of the sandwich facade, the block visual effect is also taken into consideration. In the facade, the gray mirror is also used to weaken the existence of the column and wall. The visual sandwich is like two huge boxes suspending in the air visually with freely planted plants to create an indoor air garden.

项目所在的建筑拥有落地大玻璃幕墙，通透的建筑本身让室内空间的呈现有更多的可能性。设计师把空间划分为前后两个高度不同的空间，首先进入门口的前半部分是整个空间与外部环境相互吸引的视觉亮点，利用建筑本来的天然优势前半部分设置成一个挑高的空间，呼应本店的“花园”主题选用富有自然质感的粗材——麻绳，在主入口的挑高空间和楼梯上方设计了一组用线排列组合形成的“山形”艺术装置，并从“山形”中穿插着随意的灯泡点光源，宛如蜿蜒起伏的山脉上满天繁星的美景，设计以粗材精作的理念把麻绳装置发挥出远远超出其本身价值的作用。

室内除掉管道层的最低点只有约5m的高度，在如此尴尬的高度下设计需要考虑新增夹层来满足业主的商业需求。所以在后半部分的空间设置成了两层的用餐区域，在保证一层最低点高度在2.4m的基础上，选用镜面不锈钢的材料来用作一层的天花铺贴，由于镜面不锈钢表面的不平整原理，使天花呈现出犹如水纹的梦幻效果，并且一定程度上降低了一层层高不足的压抑感。

在夹层立面上设计也略为其考虑了一下体块视觉的效果，在立面上运用灰镜来弱化柱位以及墙体的存在，视觉上夹层犹如两个巨型的方盒悬停在空中，并配以绿植随意地下随，营造了一个室内的空中花园。

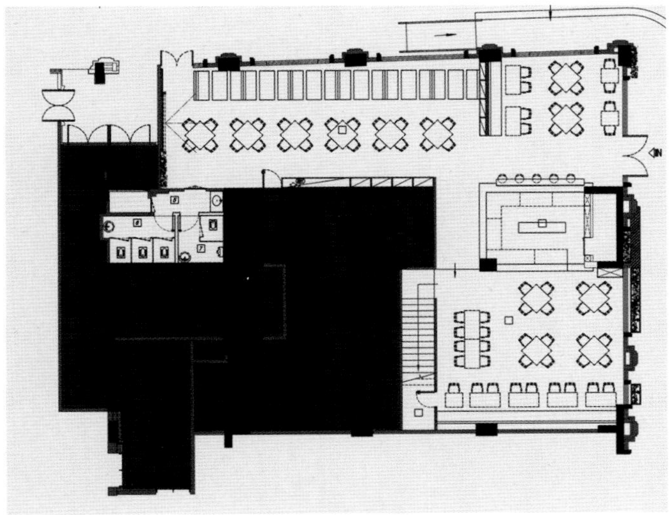

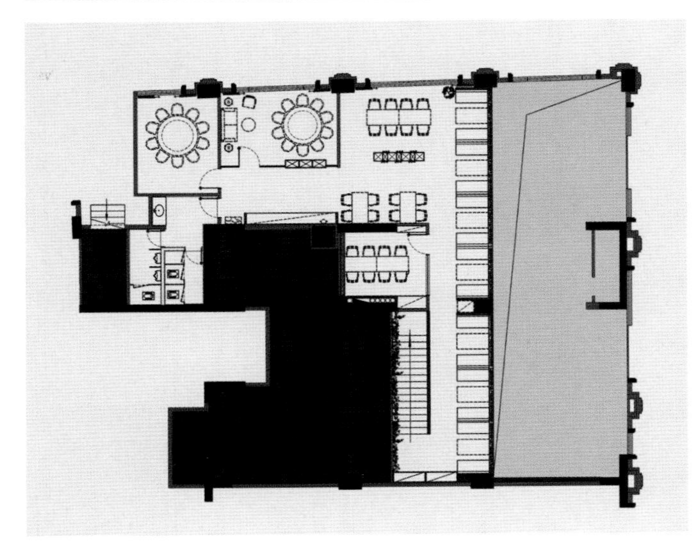

In the selection of main materials, echoing the "garden" theme, the cement paint, granite, veneer and other materials with a natural color to polish the indoor atmosphere. At the partition of the main entrance of the dining area and the stair partition, the POBO white color gradient glass is used to cover the space with a hazy haze. In the design, a silver mirror is cleverly used at the stair to extend the entire wall of green plants and "mountain shape" scenery and space. In the design, the block adding and decreasing technique is used. Materials are cut, and inset and different tiles are used in the ground paving to divide the bricks into different areas. For example, the bar seat area is divided with tiling achieving a balance between practice and aesthetics.

主材的选择上同样呼应"花园"主题采用水泥漆、麻石、木饰面等自然色系的材质来润色室内的氛围。在主入口用餐区的隔断和楼梯的隔断特意选用了POBO的白色渐变玻璃，为空间的氛围蒙上一层朦胧的薄雾。设计在楼梯处巧妙地用了一面通顶的银镜，把整面的绿植墙和"山形"装置的景象和空间感得以延伸，在吧台的设计上运用体块加减的手法，通过材质与材质间的镶嵌切割而成，在地面材质的铺贴形式上也采用不同的砖来划分出不同的区域，例如吧台座椅的区域采用花砖划分，在实用与美学之间取得了平衡。

THE EXQUISITE CUISINE, THE NATURE DECORATION

精致料理 天然雕饰

Project Location

Project Name | 项目名称
扬州虹料理

Design Company | 设计公司
上瑞元筑设计有限公司

Chief Designer | 主创设计师
孙黎明

Participant Designers | 参与方案设计
耿顺峰、胡红波、徐小安、陈浩

Area | 项目面积
580 ㎡

Photographer | 摄影师
陈铭

Main Materials | 主要材料
新古堡灰石材、酸洗锈石、波浪不锈钢板、楸香木木饰面、实木地板、草编墙纸、亚克力棒、镀铜金属件、中国黑石板等

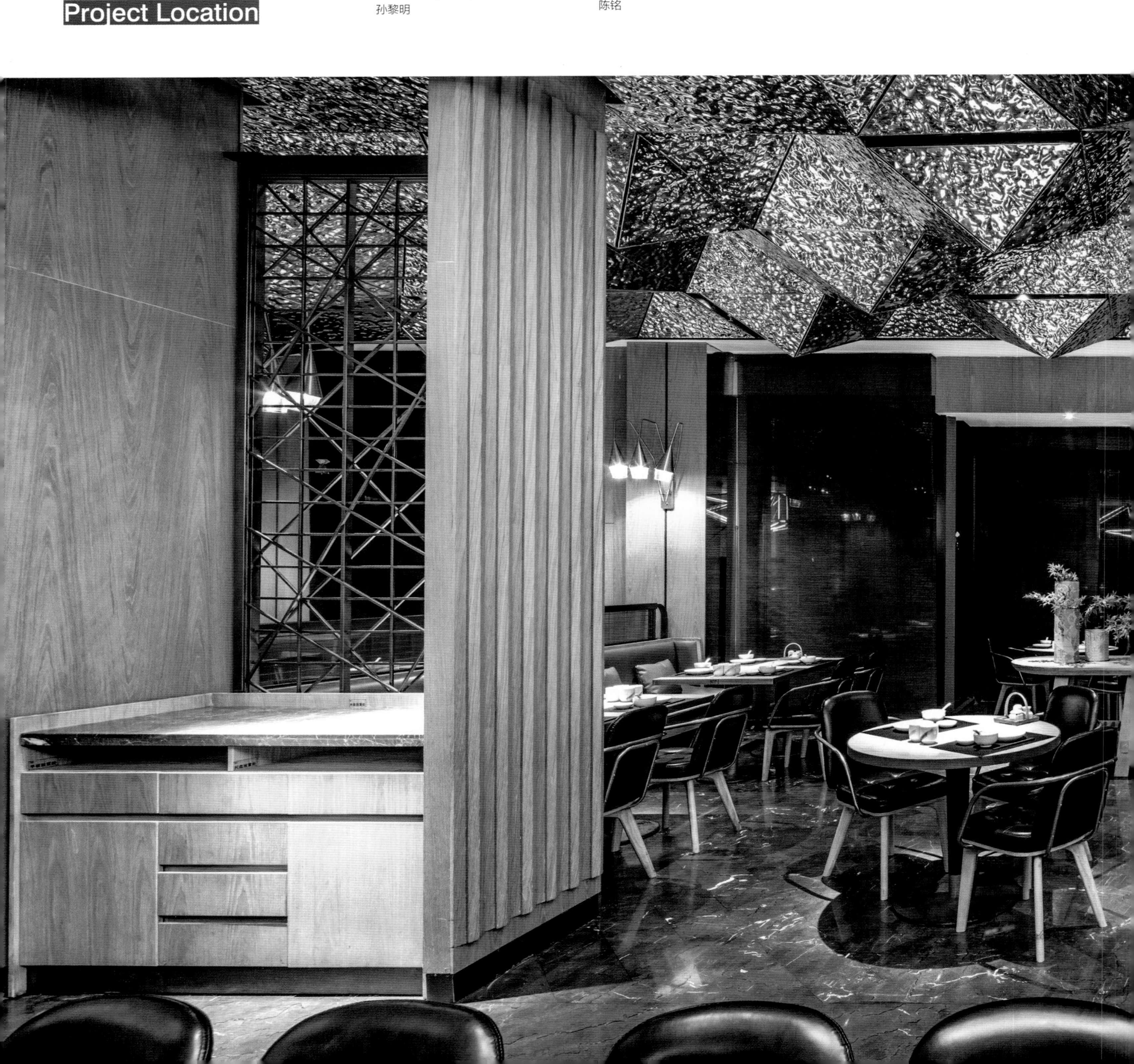

The cooking process of Japanese cuisine is the most meticulous international cuisine recognized in the world. It has no-carved natural food materials, elegant natural pottery and raw-wood tableware. One of the most worth mentioning, undoubtedly, is the dining environment noted for its elegance and simplicity, which also creates the exquisite and healthy diet concept of Japanese cuisine. HONG Cuisine also follows this concept: natural flavor, delicate and exquisite, well-made, materials and conditioning techniques paying attention to seasonal sense. This spirit has also been applied to the design of post-restaurant. In the design stage, the designers make the Japanese traditional culture performed through modern design techniques to allow the Japanese traditional culture to influence the diners imperceptibly.

日本料理拥有无雕无琢的自然食材；一丝不苟的烹调过程；素雅天然的陶器、原木食具。其中最值一提的，还当属它以古朴典雅著称的用餐环境。这也造就了日本料理精致而健康的饮食理念。扬州“虹料理”店也沿袭了这一理念：自然原味、细腻精致、制作精良，材料和调理手法重视季节感。这一精神也被运用到后期餐厅设计中，在设计阶段设计师将日本传统文化用现代设计手法加以表现，让日本传统文化在无形中影响食客。

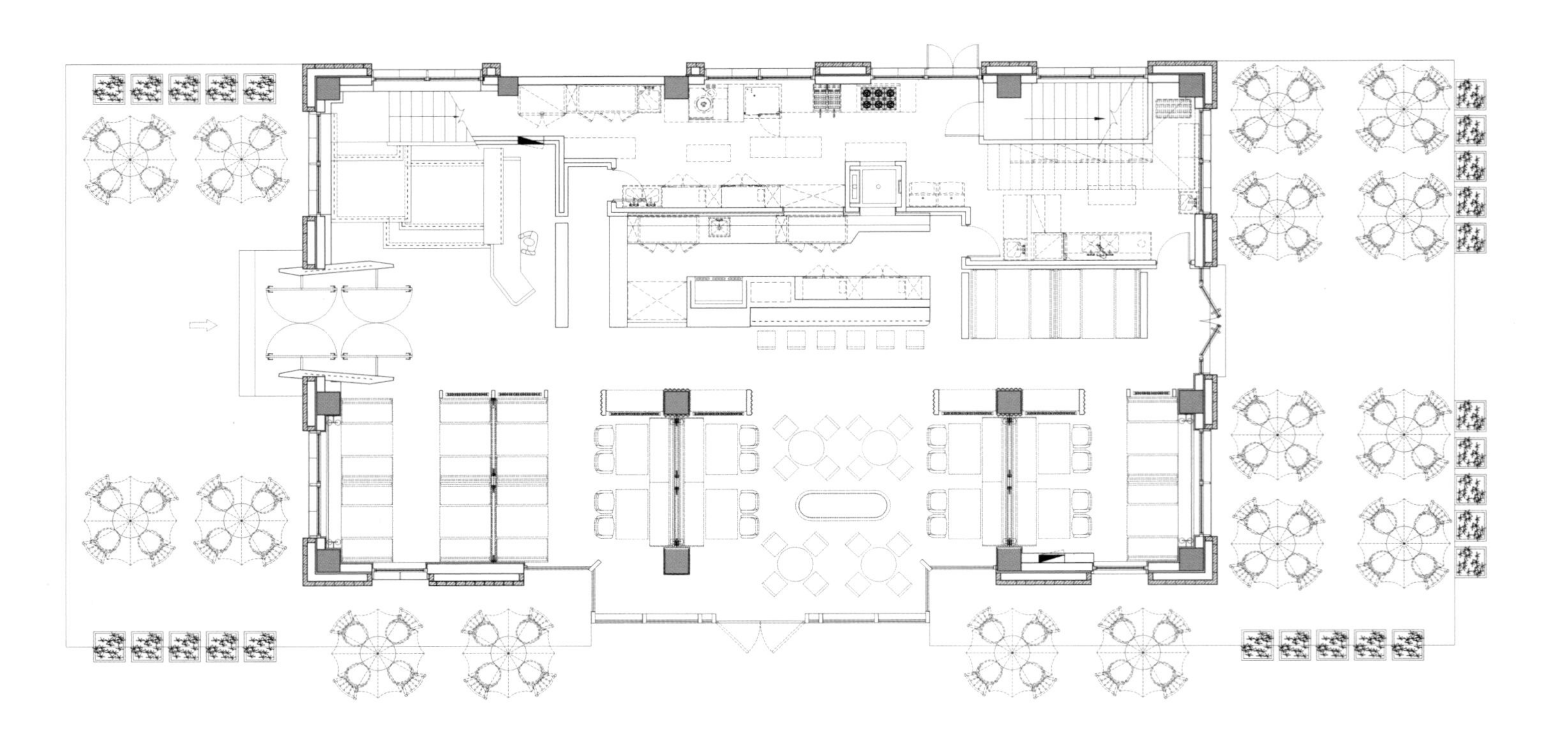

NI
HON
RI

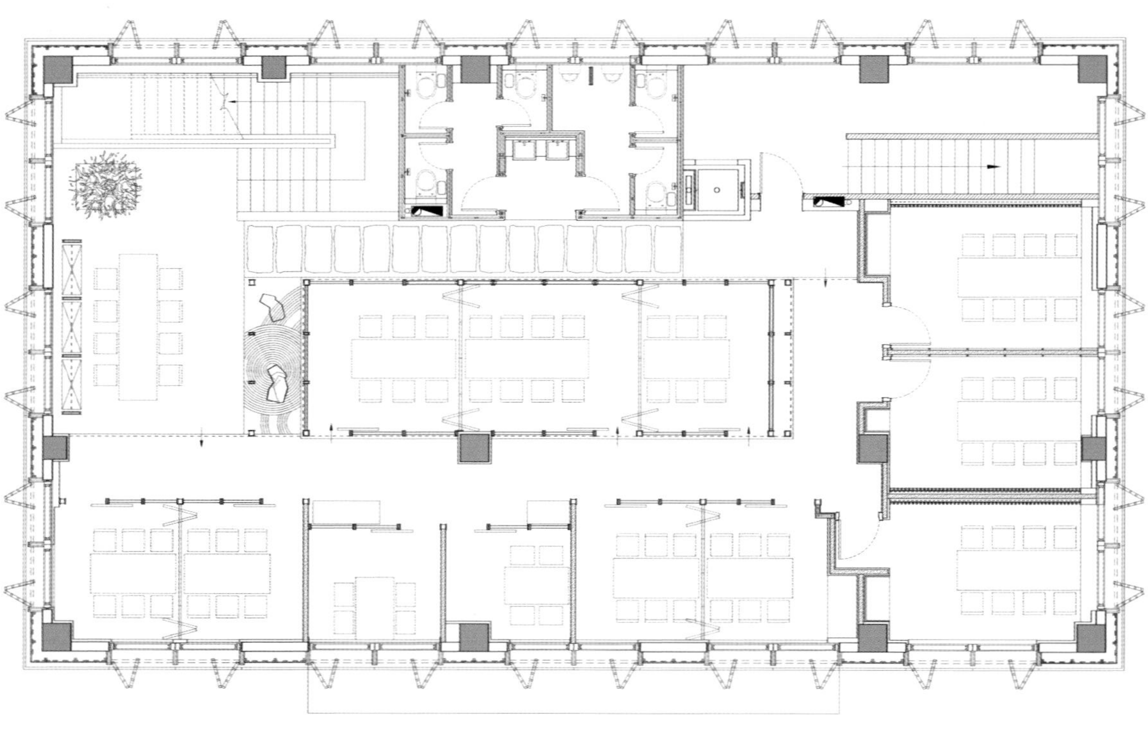

Designers use a large number of Catalpa wood veneers, straw wallpapers, pickling rust stones, new castle gray stones and other modern decorative materials for space creation. A large area of wave stainless steel plate is throughout the whole roof, which its inspiration comes from "ISSEY MIYAKE" rhombus series to deliberately build the image of glistening inverted reflection of Mount Fuji mountains in water, restrained without losing the amazing. The acrylic rods in the aisle are built into an abstract device, like a large snow floating in the air and also like flying sakura. The beauty of arrangement arises spontaneously through large and small contrast. On one side of the aisle, there is semi-hidden landscape painting in the grid; on the other side, there are geometric pickling rust stones with a large area, seeming contradictory but well reflecting the root of Japanese culture, virtually combined with the traditional Japanese elements, different from traditional design in details. Diversified design of space makes the whole environment free from vulgarity, with gentle breeze, reflecting a more layering, with infinite meaning.

设计师大量运用楸香木木饰面、草编墙纸、酸洗锈石、新古堡灰石材等现代装饰材料进行空间打造，大面积的波浪不锈钢板贯穿于整个屋顶，灵感来源于“ISSEY MIYAKE”的菱形系列，有意打造成设计师心中富士山山峦在波光粼粼的水中倒影的形象，内敛而不失惊艳之处。过道中亚克力棒则被打造成了抽象的装置，好似飘在空中的大号雪花又好似飞舞的樱花，大与小的对比，层次之美油然而生。走道一侧是栅格里若隐若现、虚虚实实的山水墨彩，另一侧则是几何分割的大面积酸洗锈石，看似矛盾却恰恰反映了日本文化的根本，无形中与传统日式元素相结合，在细节上又与传统设计有所不同。空间的多元化设计使得整个环境脱俗，和风煦语，折射出更多层次感，意味无穷。

A YOUNG PETTY-BOURGEOIS SENTIMENT

正青春的小资情调

Project Name | 项目名称
Tiyaa 西餐厅

Design Company | 设计公司
TAO studio

Designer | 设计师
刘涛、曲方圆、张晋

Soft Decoration | 软装设计师
孙晓钰

Area | 项目面积
460 ㎡

Photographer | 摄影师
孙捷先

Tiyaa is a western restaurant on the Taiwan Road, Qingdao. The owner was born in the 1980s who came back from aboard. He has a lot of ideas about his restaurant, and the most is that he wants to change the colors on the original basis.

After the analysis of the design team, designers decided to divide the restaurant into three areas with three themes: Alice in Wonderland area, warm and romantic dating area and young and passionate party area.

Entering the restaurant, there is a large vine tree climbing along the wall, red flowers in warm light appearing unlimited romance. The sight will be drawn into the hall, which is the first theme: Alice in Wonderland. Walking in it, you can see a large wall with the theme of preserved moss. Here is the perfect position for taking pictures.

Tiyaa

Tiyaa

Then it is the second theme area of the restaurant. Respecting the owner's advice, designers retain the original bookshelf wall and make a partial modification to the wall with colors. A large picture of zebra is divided into the original shape of bookshelf, adding vitality to the original dull bookshelf. A flushbonding table in bookshelf uses the brass plate and retro blue color to create a sweet and romantic dinning atmosphere where is the best choice of dating.

After 9 o'clock, with the lights gradually began weak, Tiyaa becomes another different style. The most inside rooms in restaurant are private rooms which suit for young people who are fond of party. The place which uses the fiery red can be decorated into different style according to the theme of party.

Tiyaa是位于青岛台湾路上的一家西餐店，老板是个80后，从国外回来的他对自己的店有很多想法，最想要在原有的基础改出点“颜色”瞧瞧！

在设计师团队经过分析后，决定把店划分成三个区域，分别定为三个主题：爱丽丝梦游仙境区、温馨浪漫约会区、青春热情Party区。

推开店门，迎面便是一棵大藤蔓树沿着墙面攀爬而上，红色的花在暖光中显得无限浪漫，视线随之被牵引至大厅，也是第一个主题——爱丽丝梦游仙境。走进去就看见一面大大的永生苔藓主题墙，这可是拍照的绝佳位置。

接着是餐厅第二个主题区域，尊重老板的意见，保留了原有的书柜墙，设计团队为了添加色彩，对墙体做了局部改造。一幅大斑马挂画分割成书柜原有的造型，给原本单调的书柜增添了活力，并在书柜处的嵌入式餐位用了黄铜板和复古蓝，营造了甜蜜浪漫的就餐氛围，是情侣约会的首选。

到了晚上9:00后随着灯光逐渐变弱夜场开始了，Tiyaa又是另一种不一样的格调。餐厅最里面是包房，适合喜欢Party的年轻人。整体用了热情似火的大红色，场地可以根据聚会主题来布置不一样的环境。

Get You
Hands
Dirty

Dirty

coffee
from the occident
and found a new
miracle in th
East.It is no
simple drink
1500
Get Your
Hands
Dirty

A TROPICAL PARADISE

热带天堂

Project Name | 项目名称
Ratten 餐厅

Design Company | 设计公司
MINAS KOSMIDIS (ARCHITECTURE IN CONCEPT)

Lighting Design | 灯光设计
MINAS KOSMIDIS

Area | 项目面积
325 ㎡（室内）,120 ㎡（室外）

Photographer | 摄影师
N.VAVDINOUDIS - CH.DIMITRIOU

Rattan is a bar-restaurant which is located in Faliraki of the island of Rhodes. In Faliraki, the architectural anarchy and aesthetic confusion, which are easily perceived when visited, are dominant; this fact, however, does not prevent the area of being one of the main touristic attractions of the island due to its really beautiful beaches. The location of the building was totally absent of any interesting characteristics and consequently, neither the built nor the natural environment of the wider area could have been used as a source of inspiration for the project. The restaurant actually occupies the ground floor of a building. Its only advantage is its distance from the city of Rhodes as it is relatively close (just 10 minutes by car) and as a result, the new restaurant could focus more on the permanent residents of the island without, nonetheless, excluding the tourists or the passing by potential customers.

Ratten是一家位于罗德岛法里拉基村的酒吧餐厅。在参观法里拉基村时，人们很容易感知到建筑错乱感与审美混乱感。然而，这并不能阻止该地因其美丽的海滩成为岛上一个主要的旅游区。这栋建筑完全没有位于任何一个可以带来项目灵感的具有特色的地方，也不位于有着宽阔视野的自然环境中。实际上，餐厅占据了整栋建筑的底层，它唯一的优势就是距离罗德市相对来说比较近（仅需10分钟车程）。因此，新餐馆可以更多地关注本岛居民，但不包括游客或路过的潜在顾客。

As a result, space had to be designed so that it could become a destination for the island's residents as well as a restaurant which by itself could attract tourists to its location. The main goal of the general design process was the creation of a space where its volumetric components would be a harmonious synthesis interrupted only by "green" and natural elements and had clear references to nature and specific to the tropical garden, trying to give a continuous and at the same time distinguishable flow between the interior and exterior.

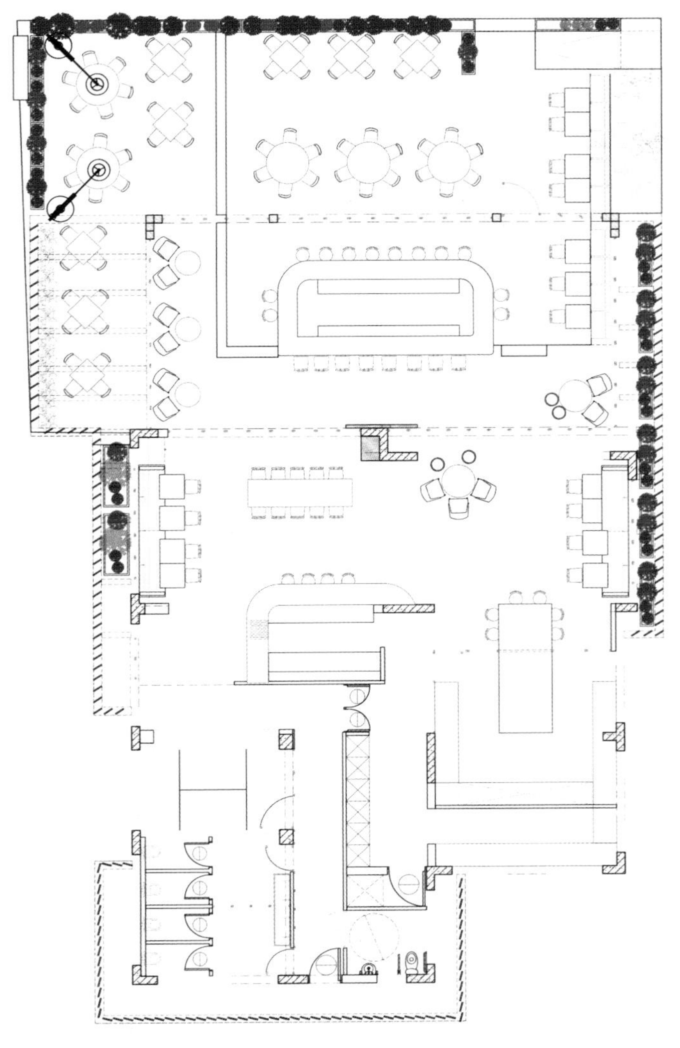

Regarding the materials and colors of the overall design, vertical and horizontal surfaces in earthy tones, with the use of wood as the dominant construction material, surround lighting fixtures and furniture made of rattan of different types and weaving, all these elements together create a composition of wood and rattan which is only interrupted by the diversity of tropical plants and colorful fabrics which create an illusion of some kind of a tropical "paradise" that charms the visitor and defines the "Rattan" restaurant's mystery.

因此，这个空间设计成为该岛居民的目的地，并且是一个本身能吸引游客的餐馆。在总体设计过程中，主要目标是创造一个其内部成分为一个仅被“绿色”和其他自然元素所打扰的可和睦共处的空间。它对自然有着清晰的认识，特别是热带花园，试图提供一个连续的同时可区分室内与室外区域的地方。

关于整体设计的材料和颜色，设计师使用土色调的垂直和水平表面，搭配使用作为主要建筑材料的木材，环绕照明设备和由不同类型和编织藤制成的家具共同创造了一个木和藤组合，被多样性的热带植物和丰富多彩的布料所阻断，可以营造出一种热带 “天堂”的气氛，吸引游客并保留了“Ratten” 餐厅的神秘。

FRAGRANCE AROUND THE TIP OF TONGUE, ENCOUNTER THE JUNGLE

舌尖绕香 邂逅丛林

Project Name | 项目名称
遇 · 咖啡

Design Company | 设计公司
北京海岸设计

Designer | 设计师
郭准

Area | 项目面积
1720 ㎡

Main Materials | 主要材料
玻璃、混凝土、钢、木、砖、石等

"She walks in beauty, like the night; Of cloudless climes and starry skies; And all that's best of dark and bright; Meet in her aspect and her eyes; Thus mellowed to that tender light; Which heaven to gaudy day denies; One shade the more, one ray the less; Had half impaired the nameless grace..." The romance is deep in Byron's heart, and he described the beautiful scenes and belles with poetry. Following his inspiration, the designer uses the original-basis as a concept to present the "dreamland" to the public and combines glass, concrete and steel with light to create an extremly elegant and warm aesthetic space.

"她走在美的光彩中，像夜晚/皎洁无云而且繁星漫天/明与暗的最美妙的色泽/在她的仪容和秋波里呈现/耀目的白天只嫌光太强/它比那光亮柔和而幽暗/增加或减少一份明与暗/就会损害这难言的美……"拜伦的血液中流淌着浪漫，他用诗歌描绘了美轮美奂的景象与伊人。追寻着诗歌带来的灵感，设计师以归本主义为理念，将这"梦境"呈现于世人，用光线结合玻璃、混凝土与钢等材料，打造出极致优雅、温馨的唯美空间。

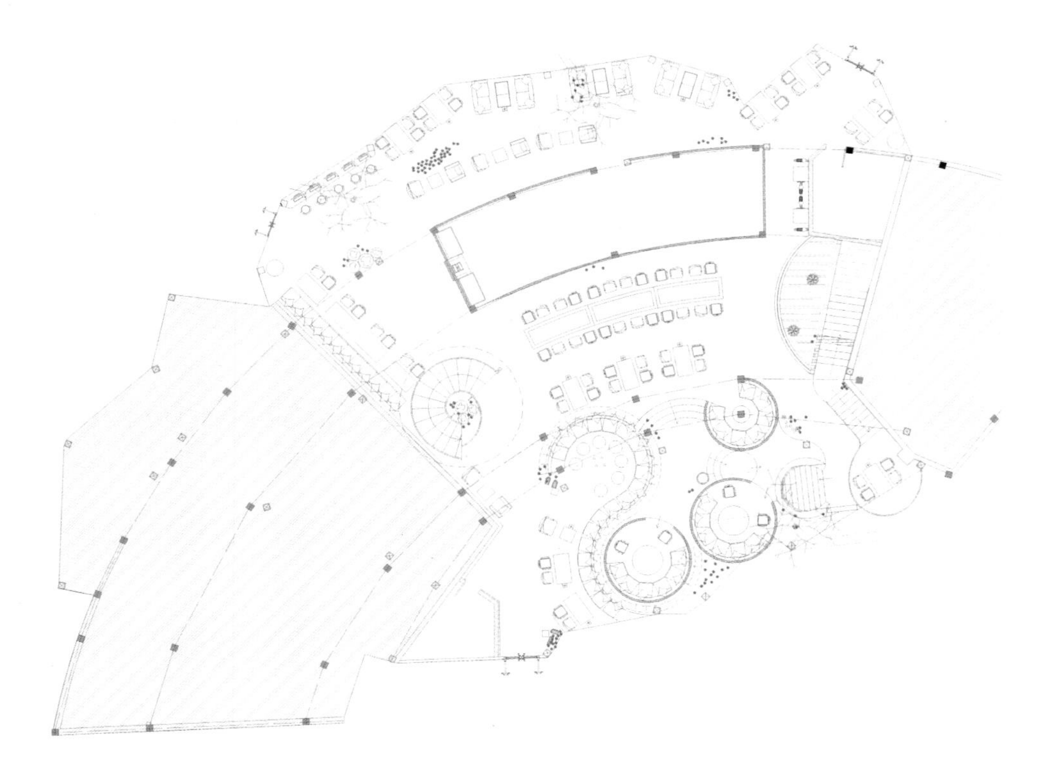

The complicated beauty of different elements, the cement ground, wood furniture and wood decorations exuding a natural flavor, and the unique crystal chandelier is like a retro fairy tale world. The rare height of space and the panoramic French windows extend a broad vision. With the delicate tree decorations, customers can breathe the fresh air and enjoy the comfortable environment. There are bookshelves, slides and lounge chairs in the rest area, customers can enjoy a lazy and leisure time here and learn to get along with the world gently.

散发着自然风味的水泥地面、木质桌椅、木片装饰，风格独特的水晶吊灯……不同元素的繁复美感交织着，像一个复古的童话世界。空间的层高罕见，设全景落地窗，具有开阔的视野。伴着随性大气的树木装饰，呼吸着清新的空气，环境极其舒适。休息区设有书架、滑梯、躺椅，在这里享受慵懒和悠闲，学会与这个世界温柔相处。

ICBC 中国工商银行

This is a two-storey space, which the first floor is used for entertaining customers, and the second floor is for children to play. Wandering in this space, customers can enjoy the beautiful scenery like trees and stones, steel and cement, mottled wall and plain tables and chairs, abundant vegetation and dim light, full of the atmosphere of art and forest. As the designer Mr. Guo Zhun said, "We hope it is elegant, generous and natural without any pretentiousness and decoration." While the scattered plants add a touch of fresh feeling to this space.

From dawn to dusk, the interaction between lights and shadows keeping changes brings the endless moving and surprises to customers. The night falls, and the gentle light enveloping the whole space as the embellishment of the light on the roof shines in the darkness, and introduces people into the infinite reverie of a starry night.

"Encounter" represents a kind of wisdom, living in the moment, and lets the romance be what it should be. Entering here, it is like walking in the jungle and leisurely in the mountains. "But it was not the wine that attracted him to this spot; it was the charming scenery." What people enjoy not only are the touches on the tongue but also let the blundering mood into calm in a comfortable and relaxed atmosphere to feel an elegant flavor and experience a beautiful encounter.

空间共分为两层，底层用来招待顾客，二层则设置为供儿童娱乐的区间，使得孩子们也可以在这里尽情玩耍。徜徉于此，树与石，钢与水泥，斑驳的砖墙与朴拙的桌椅，盎然的草木与昏黄的灯影……正如设计师郭准先生所言，“我们希望它优雅大方自然，不带任何的矫情和修饰”，文艺与森林的气息弥漫生香。而散落各处的绿植，又为画面增添几分清新之感。

从清晨到日暮，光与影相互交汇，不停演绎和变化，为顾客带去不尽的感动与惊喜。夜色降临，轻柔的光笼罩着整个空间，点缀于屋顶的明灯，在幽暗中散发着光芒，将人引入星夜的无限遐想之中。

“遇”代表着一种智慧，活在当下，让浪漫顺其自然。走进这里，犹如置身于丛林之中，悠然地在山间漫步。“醉翁之意不在酒，在乎山水之间也。”人们享受的不仅仅是舌尖上的触动，更是在舒适清爽的氛围中，让浮躁的心绪安定下来，感受一番优雅的情调，邂逅美好。

MEET CAFE
WELCOME TO LONDON
MEXICAN FOOD
Tea Cakes
38
Hackney Dalston
Islington Rosebery Ave
Bloomsbury Piccadilly
CLAPTON POND
KICK START!

MEET CAFE

QUIETNESS OF ZEN AND FLUID

禅谧 · 流体

Project Name | 项目名称
九川堂锅物 · 新北市芦洲店

Design Company | 设计公司
周易设计工作室

Chief Designer | 主持设计
周易

Participant Designers | 参与设计
陈昱玮、张育诚

Area | 项目面积
626 ㎡

Main Materials | 主要材料
黑板石、橡木皮染黑、钢刷梧桐木皮、南非紫檀木、美桧等

Photographer | 摄影师
和风摄影

"JIU CHUAN TANG", a square and refine building body with iron wooden grid exterior, through the dense vertical and horizontal line dislocation, creates a visual sense of penetration. Progressive lighting of the building body contrasts the powerful light words of shop signage, which its stable and light stature, quiet and refined atmosphere is very pleasing and charming, and slowly elaborate the beauty of Zen rhyme which couple the hardness with softness.

On the first floor, the counter uses chisel stone skin coupled with log surface, allocating with Chinese window of back walls, producing a harmonious contrast with rugged and delicate feeling. Setting a cyclic landscaped pool in the middle of the space, the edge separates the pool body and floor through a light belt, showing a sense of floating lightness. Upper side, ceiling uses a black grilled glass surface, embedded a lot of wooden structures, creating an elevation vision of "A large quantity is beautiful". The booth area sequentially arranged around the landscape pool is in an elegant style, and the rear is decorating glass cast the light on the wall and booths interpreting the silhouette artistic with scattered grids.

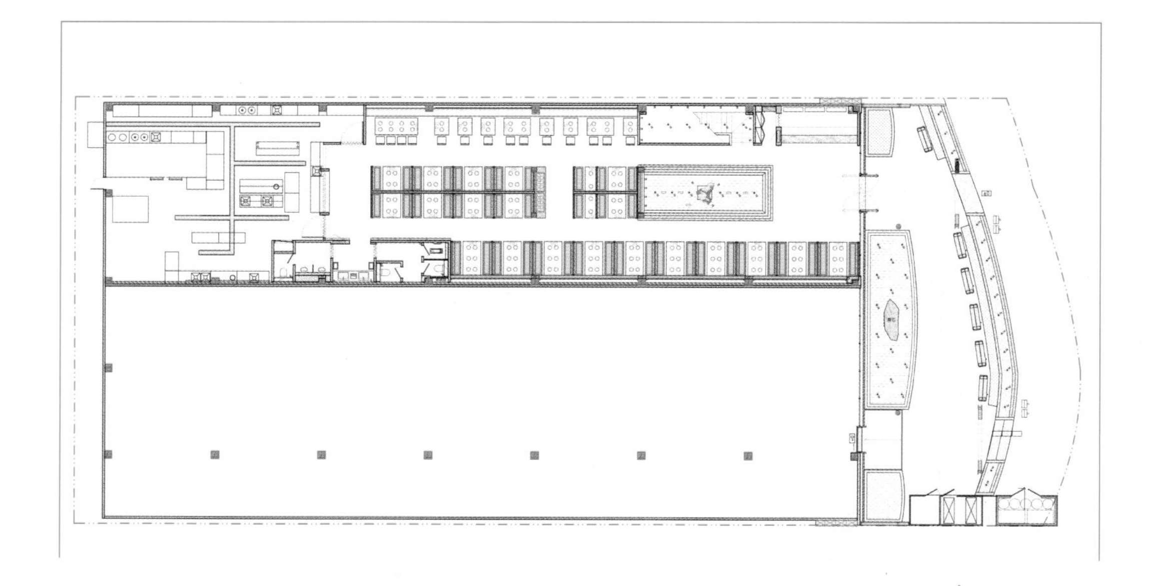

The second floor keeps and continues the landscaped pool to open the show. Huge wave grids are meticulously arranged on the top of indoor, and the diversity liquid fluid shapes the fantasy degeberation. This case combines characteristic materials and modern techniques to explain abstract Zen and interpret a new sensory experience.

九川堂，方整却精致的建筑量体，外观铁木格栅，透过疏密的直列错位，创造视觉穿透感。量体上下渐进式照明，烘托遒劲的光体字店招，其安定轻盈的身形，脱俗静谧的氛围，十分悦目迷人，缓缓阐述刚柔并济的禅韵之美。

一楼柜台以凿面石皮佐以原木台面，搭配后墙中式窗花发想，产生粗犷与细致的和谐对比。空间居中设置循环景池，边缘借光带将池体与地坪脱开，呈现轻盈漂浮感。上方天花板使用黑烤玻璃面，嵌以量取胜的木构件群落，造就仰角视野“数大便是美”的磅礴。环绕景池排序的卡座群，造型朴雅，后衬玻璃投光于墙面与卡座，以错落格栅诠释剪影艺术。

二楼延续景池开场，室内天顶缜密排列巨大波型格栅，气象万千的凝态流体型塑奇幻异次元。全案揉合特色素材与现代手法解译抽象禅风，演绎全新感官体验。

ENCOUNTER THE TASTE

味见

Project Name | 项目名称
雪月花店

Design Company | 设计公司
上海黑泡泡建筑装饰设计工程有限公司

Chief Designer | 主案设计师
孙天文

Designer | 设计师
孙天文

Area | 项目面积
1300 m²

Main Materials | 主要材料
涂料、玻璃、榻榻米等

Photographer | 摄影师
张静

No matter how concept and definition are elaborate and mellow, finally most of them will become the punishment of the "form". So, the designer Sun Tianwen said, "We can reject any form of concepts and ideas, but cannot deny the potential influence of the building and the room that you live in. In comparison, it seems that 'conveying something' is more meaningful than arguing 'what is it?'."

This kind of seemly arrogant description, in fact, is a kind of spiritual attitude full of sincerity, which is the origin of the designer transforming this restaurant's name into "Snow Flower" at first.

无论多么精巧圆熟的概念与定义，最后大都会沦为对“形式”的诛伐。所以本案设计师孙天文先生说：我们可以拒绝任何形式的理论和观点，但却无法拒绝来自其生存的建筑、居住的室内所带给你的潜在影响，相比之下似乎“传达什么”比争论“这是什么”更有意义。

这种似乎孤傲的夫子自道说到底其实是一种诚意满满的精神姿态，这也是设计师第一步就将这间料理店的名字改为“雪月花”的缘起。

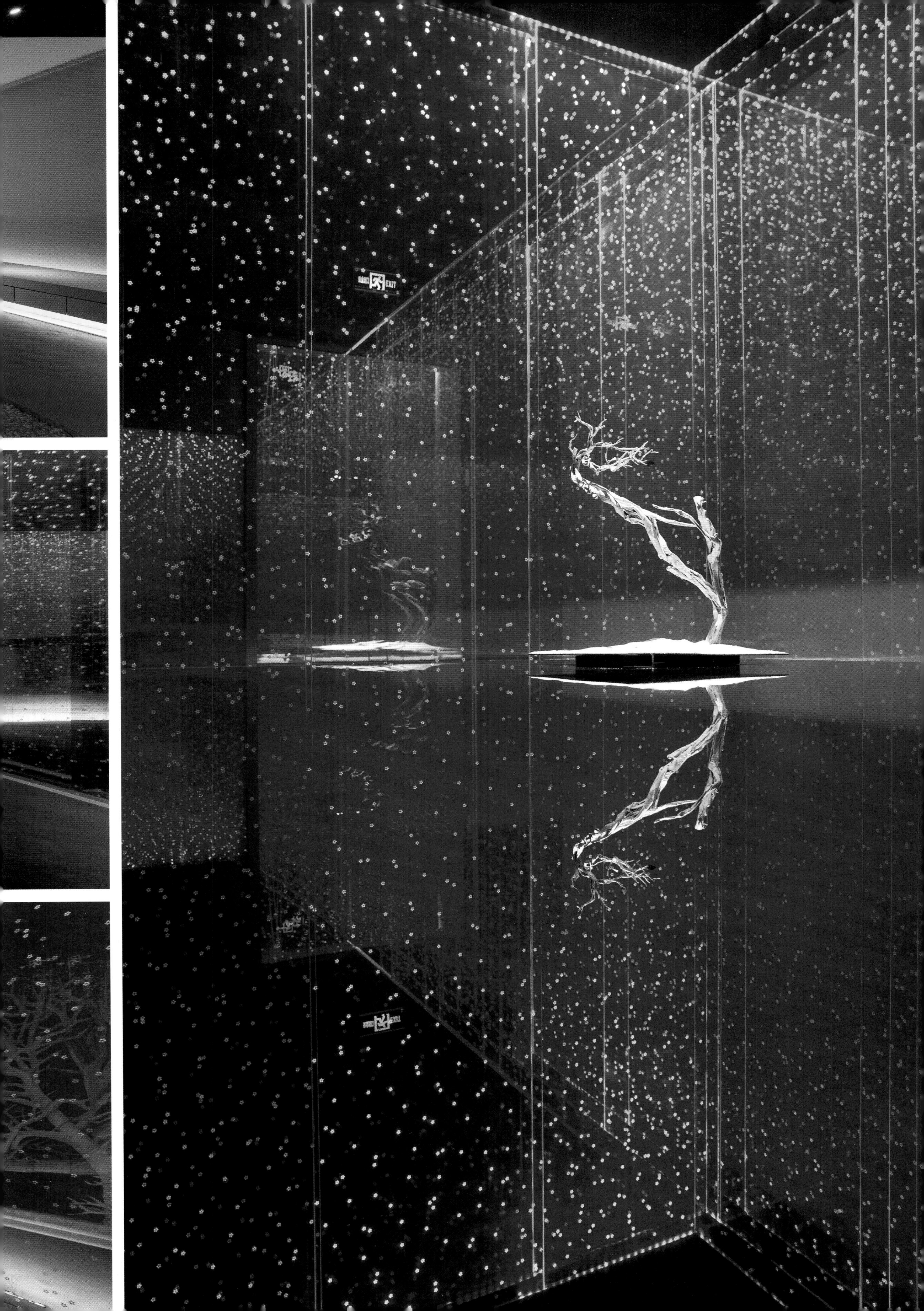
EXIT

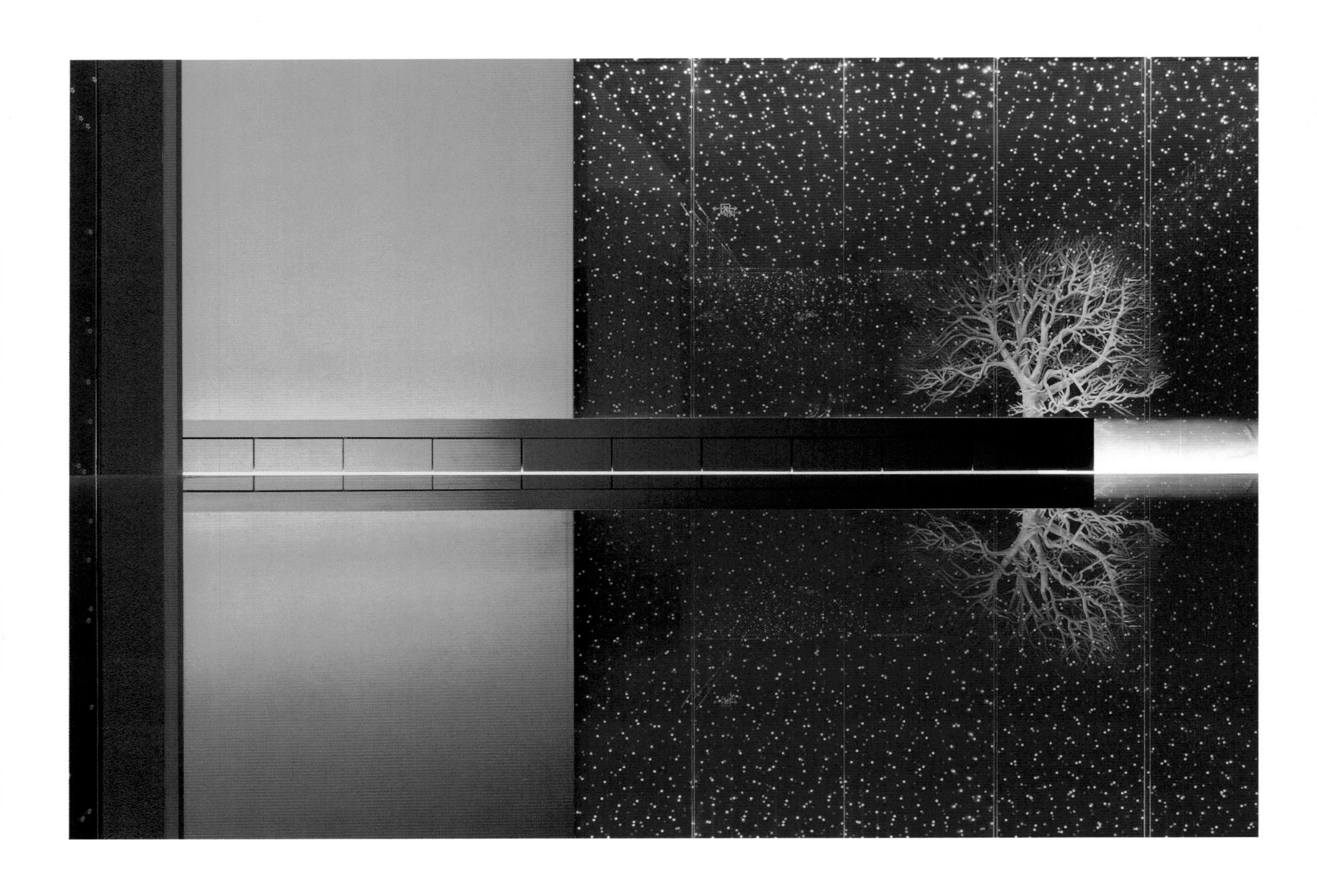

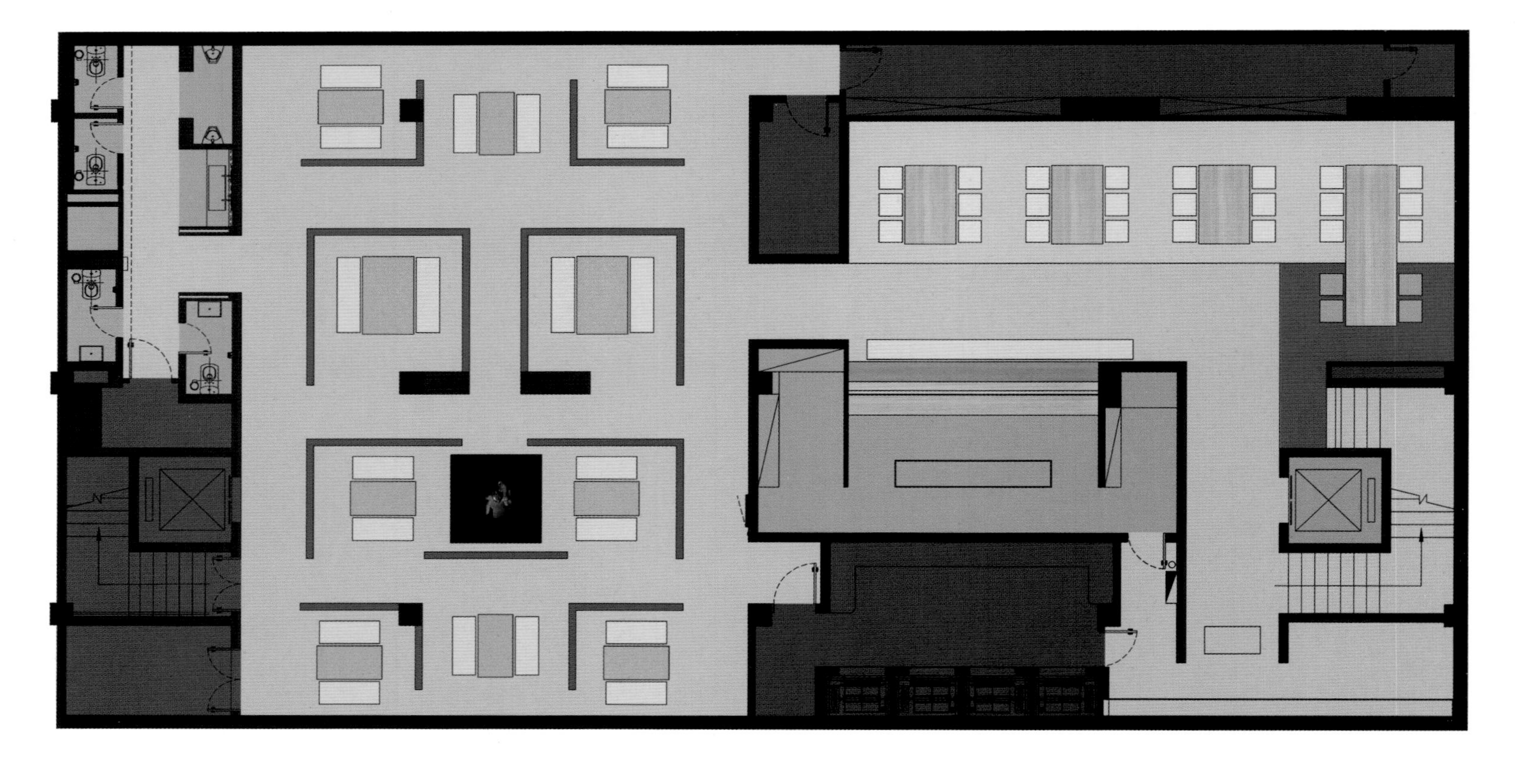

In Japanese Waka and Haiku, snow flower represents everything in nature, joys and sorrows. For the elite of interior design, the perfect technology is being part of him, so what he pursues are the delicate details of culture and heritage.

The ultra white glass carved sakura, the blue LED light band, and the black sushi table background, everywhere of these elements is stunning, but it is not sufficient to cover the overall Zen flavor and purpose. Only when these are in complete harmony to form a proper, using "proper" to remind himself to convey sincerely is the perfect cause and effect.

The essence of Japanese cuisine is "eating something in their best time" and "when the weather becomes cold, it has started to prepare for the bream", Hirohisa Koyama said. Except for the cutting skills with hundreds of practices and the precise feelings, the key of Japanese cuisine is the perception of heart. So the design does—but its fundamental is the experiences in design for many years.

无论是被称为“永远的旅人”的松尾芭蕉的一首：“今夜雪纷纷，许是有人过箱根”；还是明惠上人的“更怜风雪漫月身”；抑或是“喜见雪朝来”“花不为伊开”和“月明堪久赏”，都显示出在日本的和歌俳句中，雪月花代表了自然万物，也代表着欢喜哀愁——对于空间设计的顶尖高手而言，技术上的完美已经是题中应有之义，所争的分毫就在于文化视野和底蕴的大巧不工。

雕刻樱花的超白玻璃、蓝色的LED光带、全黑的寿司台背景……每一处都有惊艳，但每一处又并不足以涵盖整体的禅意妙旨——而只有当这些水乳交融才形成适当，以“适当”这两字提醒自己诚实地传达，便是至臻完美的因果。

日本料理的精髓是“不时不食”，“寒意上心头时，就得为鲷鱼开始做准备”，日料大师小山裕久如是说。百炼刀工、精确手感之外，关键是心的感悟，设计其实同样如此——鱼跃在花见，花开在眼前，用刹那，问候浮生，而其根本却端在于浸淫行业多年的修为。

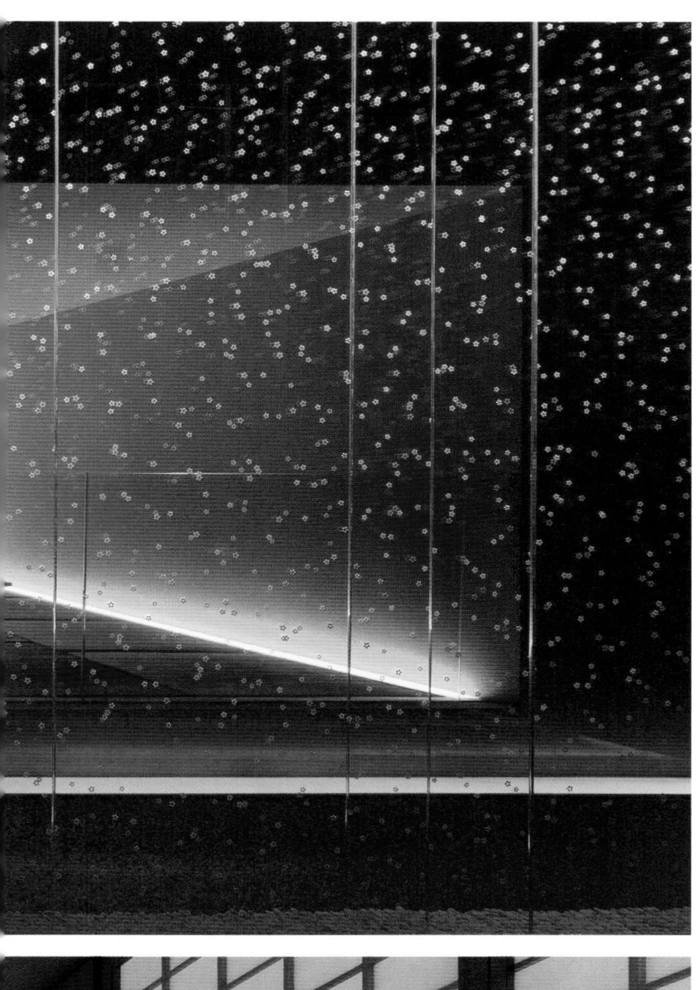

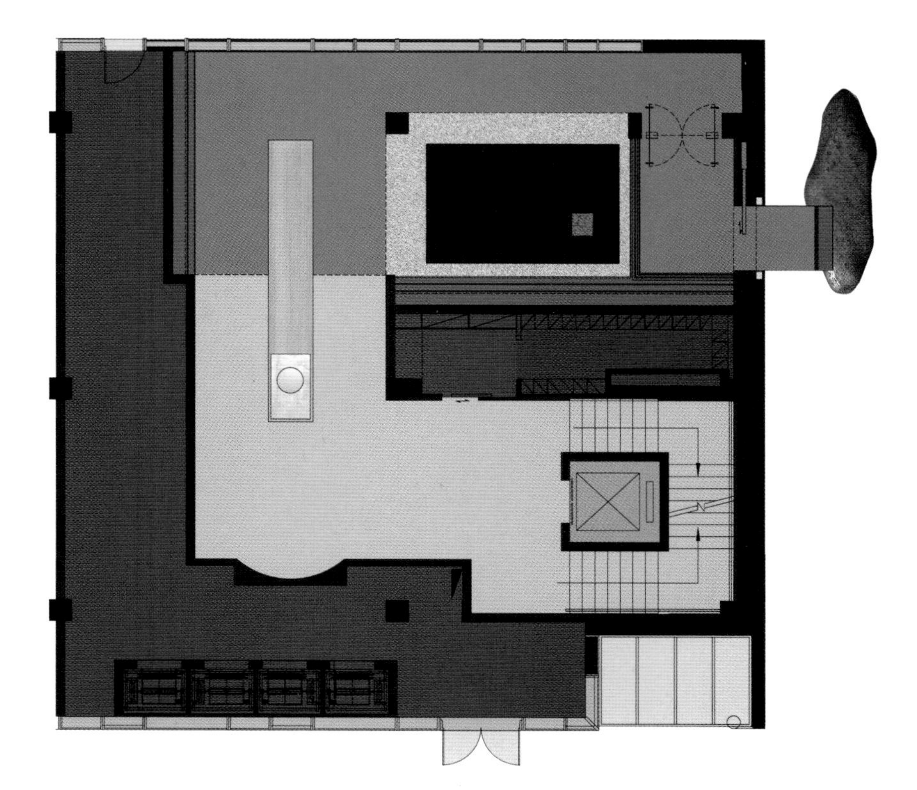

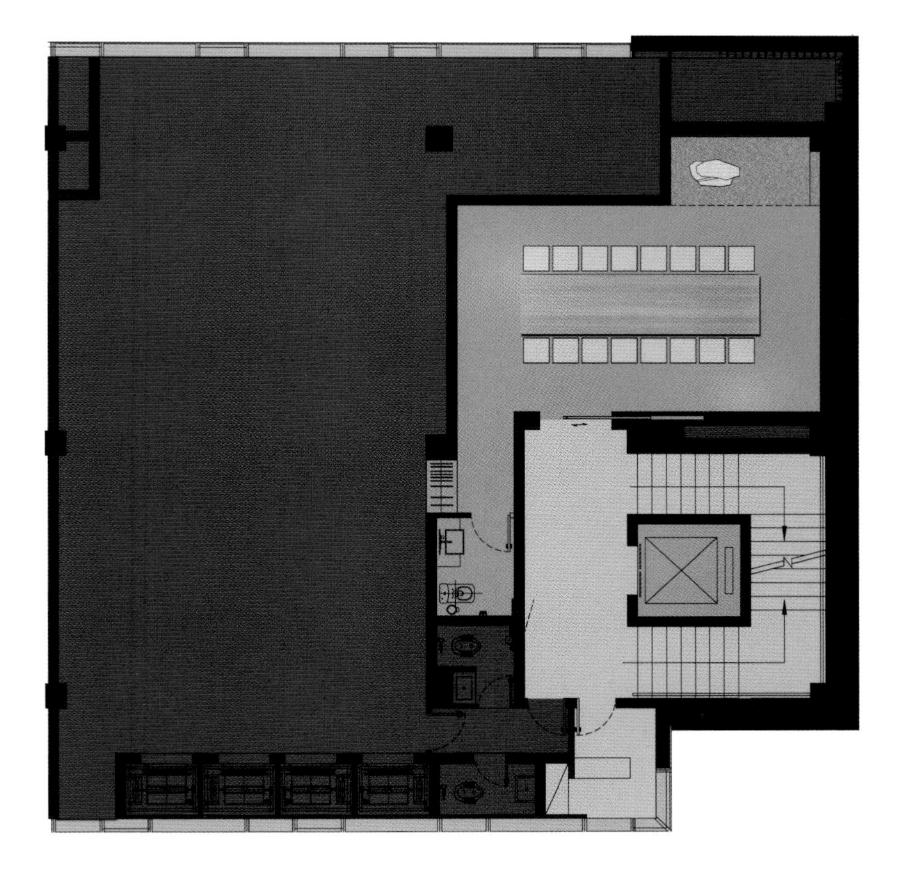

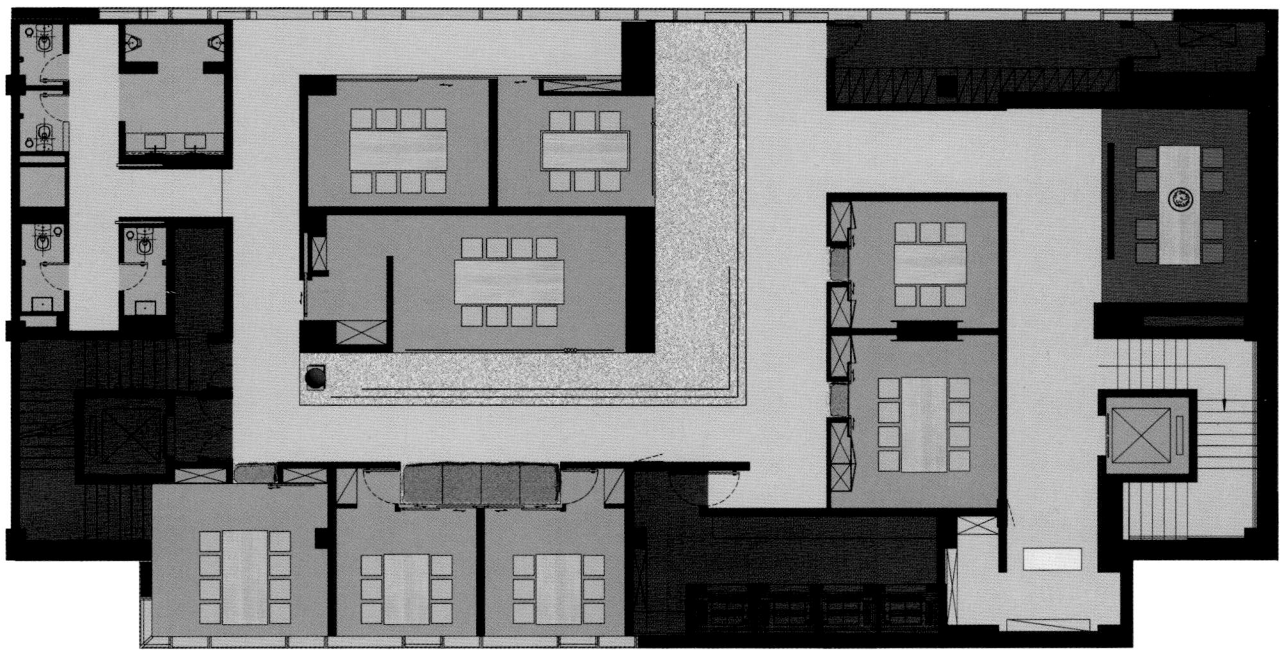

图书在版编目（CIP）数据

欧陆食代：国际创意餐厅设计. II / 深圳视界文化传播有限公司编. -- 北京：中国林业出版社，2018.2
ISBN 978-7-5038-9416-9

Ⅰ. ①欧… Ⅱ. ①深… Ⅲ. ①餐馆－室内装饰设计 Ⅳ. ① TU247.3

中国版本图书馆 CIP 数据核字（2018）第 017940 号

编委会成员名单
策划制作：深圳视界文化传播有限公司（www.dvip-sz.com）
总 策 划：万 晶
编　　辑：杨珍琼
校　　对：孙神英 尹丽斯 陈劳平
翻　　译：侯佳珍
装帧设计：叶一斌
联系电话：0755-82834960

中国林业出版社 · 建筑分社
策　　划：纪 亮
责任编辑：纪 亮 王思源

出版：中国林业出版社
（100009 北京西城区德内大街刘海胡同 7 号）
http://lycb.forestry.gov.cn/
电话：（010）8314 3518
发行：中国林业出版社
印刷：深圳市雅仕达印务有限公司
版次：2018 年 2 月第 1 版
印次：2018 年 2 月第 1 次
开本：235mm×335mm，1/16
印张：20
字数：300 千字
定价：428.00 元（USD 86.00）